MANAGEMENT CAREERS AND EDUCATION IN SHIPPING AND LOGISTICS

. and Education
stics

g

xon OX14 4RN
USA

ancis Group, an informa business

LC control number: 00134477

Contents

List of Figures

List of Tables

List of Appendices

Acknowledgements

My thanks are due to many colleagues who have assisted in this work. In particular the patience of Marie Bendell in typing and checking the draft manuscripts is noteworthy. Working through chapter by chapter, only a few of the others who have given of their time, resources and support without any monetary reward can be named.

Professor John Hibbs having patiently read through advanced versions of the manuscript, kindly made useful suggestions for improvements.

Peter Holgate of the Seale Hayne Faculty of Agriculture, Food and Land Use at the University of Plymouth very kindly administered questionnaires to finalist Rural Resources students that contributed to the analysis of perceptions of the role of the Transport Planner reported in Chapter 3. Dr Harry Heijveld of the Institute of Marine Studies at the University of Plymouth acted as 'Researcher B' in this context, conducting a blind content analysis of initial replies, and assisting with some of the statistical analysis involved in classifying student responses.

Thanks are due to editors and anonymous reviewers of work that has already been published elsewhere. In particular Dr Jerry Wilson of the Georgia Southern University, editor of the *Journal of Transportation Management* made very helpful comments in developed and publishing the section of Chapter 3 which discusses changing perceptions of the role of the logistics analyst. Dr David Menachof assisted by presenting an early version of this paper at a conference of logistics educators in Atlanta, returning with useful peer comment.

Within the Institute of Marine Studies at the University of Plymouth, successive course co-ordinators of postgraduate programmes have assisted me in interviewing and surveying three cohorts of postgraduate students, without which Chapters 4 to 7 would not have been possible. Thanks are due to Drs Keith Mason, Richard Gray and David Menachof in this context. The patience of students who participated in data collection sessions is noted, especially discussions such as those recorded in Appendix 1.

Colleagues in the Department of Psychology at the University of Plymouth have assisted with the research design and the early formulation of questionnaires. In particular much of the research design in Chapters 4 and 5 flowed from discussions with Professor Steve Newstead and Dr Arlene Franklyn-Stokes. David Spicer of the University of Plymouth Business School provided useful ideas on the application of cognitive mapping techniques in Chapter 6.

Colleagues nationally within educational institutions in the UK in departments of shipping, logistics and related studies have contributed to the process of collecting the data reported in Chapter 8. These included Captain Steve Bonsall at Liverpool John Moores University, Paul Fawcett and Drs William Murray and Nick Hubbard at the Universities of Salford and Huddersfield, Dr Peter Marlow at the University of Wales in Cardiff, Drs Mervyn Rawlinson and David Glenn at London Guildhall University, and Christopher Atwell at the Southampton Institute of HE at Warsach in Southampton. Without their assistance, data collection would have been much more troublesome.

The work on Port State Control and Port Logistics in Chapter 9 summarises a survey undertaken by Hicham Elmdaghri as part of his Masters thesis whilst a postgraduate student at Plymouth. Any errors in the updates, write-up of systems theory and proposed model of an enhanced training and education system are however the sole responsibility of the author.

My thanks are due to the University of Plymouth for funding attendance and presentations at a series of conferences, which enabled ideas to be aired and useful suggestions for developing work to be received from peers. In addition to those mentioned above, presentations were made at the:

UK Council for Graduate Education, Summer Conference, Leeds, July 1999 (parts of Chapters 4 and 7);
Institute of Mathematics and Its Application, Conference on Mathematics in Transport Planning and Control, Cardiff, April 1998 (early part of Chapter 3);
Computers in Psychology, York University, April 1998 (Chapter 6);
CTI (Accounts, Finance and Management), Brighton, April 1999, where the methodology discussed in Chapter 5 was aired;
The International Congress on Maritime Technological and Research Innovations in Barcelona, April 1999, where the latter part of Chapter 7 was debated.

1 Human Resource Availability in Shipping and Logistics

Background

The need to recruit and develop high calibre potential managers capable of shaping the future is a major concern in both shipping and logistics:

> It can be confidently concluded that the main problem the shipping industry is facing, or will face, is not the number but the quality of crew. Seafarers who can competently man different ships, whatever their type, size, ownership or management, with a good command of English and skill in communicating with colleagues and managers from different cultures and backgrounds, will continue to be in great demand (Li and Wonham, 1999).

> There are a number of reasons for studying recruitment issues in logistics.
> - Current skills shortages in logistics at all levels
> - The public image of logistics is generally a poor one… an industry with an unpleasant working environment and anti-social shift patterns.
> - …There is still a limited awareness amongst school leavers and many graduates of what logistics is and the range of jobs its involves.
>
> (Mills et al, 1999, p.1)

Considering initially international shipping, three issues emerge:

1. What is the quantitative evidence of changes in the supply of, and demand for particular employment groups, for example deck officers?
2. What policies might the industry, or individual organisations within it adopt to address the situation?
3. What might the impact of these policies and trends be on the individual employees in relevant occupations? This issue implies both a need to raise the competencies of human resource bases through advanced education and training, and to understand how individual managers might be enticed to partake in advanced study. It shapes much of the subject matter of this volume.

Table 1.1 Employees in employment, number of ships registered and number of deck officers in the UK (1973-1998)

Year	Employees in employment (000)	UK registered fleet - number	UK deck officers (000)
1973	86	1776	
1975	85	1682	
1977	80	1545	38
1979	74	1305	33
1981	66	1118	29
1983	49	866	21
1985	35	693	14
1987	33	506	10
1989	33	450	8
1991	32 / 36	409	9
1993	22	344	9
1995	25	365	8
1996	23	377	
1997	23	392	
1998	21		

Sources: Employment:
1973-1982: DTp (1984), Table 1.25, p.40. Standard Industrial Classification (1980), Class 74: Sea Transport. Employees in employment at June.
1984-1991: DTp (1993) Table 1.26. Employment on transport, p.53. Standard Industrial Classification (1980), Class 74: Sea Transport. Employees in employment at September.
1991-1998: DETR (1998a) Table 1.23. Employees in employment in transport and related industries: March 1988-1998, p.49. Standard Industrial Classification (1992), Class 61: Water Transport.
Registered fleet: 1973-1983 DTp (1984), Table 4.16. p.141.
1986-1997: DETR, (1998a) Table 6.15. p.145.
1977-1987: UK shipping industry. DTp, (1988), Table 4.24, p.187. Note this period excludes cadet officers.
1988-1995: DETR, 1997.

Quantitative Changes in the Supply of and Demand for Deck Officers

The 'numbers issue' is pressing and inescapable. In the case of the UK, it has been claimed that the declining numbers of qualified and trained deck officers has reached critical manning levels (Moreby and Springett, 1990) threatening the very existence of the UK's long term claim to being a nation of seafarers. In terms of the number of employees employed in the UK fleet, a 75 per cent decline in two decades (Table 1.1) from around 80,000 seafarers in 1977 to around 20,000 in 1997 has been catastrophic. This decline is closely matched by a similar fall in the number of vessels in the UK registered fleet, from around 1600 in the mid-1970s to under 400 vessels in the mid-1990s. Overtly, this trend has had an obvious impact on directly reducing the number of opportunities available to qualified and experienced seafarers, with all the consequent impacts on morale within the profession. Less overtly, it has also had a more damaging and subtle effect in helping to create and fuel a stereotype of a sector in decline, offering poor long term prospects and a diminished social and economic status to potential recruits. The number of UK deck officers has similarly waned, registering a 75 per cent decline in little over a decade. In the best traditions of a reactive 'fire-fighting' crisis management response to this decline, several inquests have sought to determine the causes of such a rapid reversal. One notable study attempted to establish a methodology for projecting the impact of a range of sea cadet recruitment policies on future deck officer manning levels (McConville et al, 1998), but nonetheless produced equally gloomy findings. Not only have existing manning levels shown to be in decline, but rises in the average age of the stock of experienced deck officers have promulgated further anticipated crises occasioned by pending retirements within the next two decades.

Fleets other than those flying the UK flag have suffered from a decline in their registered number of vessels, with similar trends also apparent in several other advanced economies (see Table 1.2). In particular, declines of around 50 per cent have been recorded in the USA, the Netherlands, and West Germany, with lesser declines in France, Denmark and Portugal. The dynamics of an overall growth of 25 per cent in the world fleet have not only benefited flag-of-convenience states, such as Panama, but also some Mediterranean states such as Italy and Greece which have maintained their overall magnitude, despite volatile fluctuations year-on-year. Mitigating explanations of these trends have included:

1. A rise in the overall size of ships.
2. Increasing ship operating speeds and productivity.

Both effects may have helped to account for the reduced requirements for large fleets and hence reduced manning needs in some economies.

Table 1.2 Number of vessels in national merchant fleets

Flag	1970	1975	1980	1985	1990
EUR12	11105	11044	11023	8552	6370
West Germany	2409	1578	1491	1447	880
Denmark	874	950	746	607	570
Greece	1604	2561	3634	2353	1520
France	514	562	469	381	280
Italy	1127	1222	1158	956	1000
Netherlands	1173	802	655	630	550
Portugal	152	169	118	112	90
United Kingdom	2355	2246	1780	1135	790
Spain	755	804	805	740	500
Norway	1860	1887	1537	1319	1680
Liberia	1823	2473	2337	1722	1610
Panama	782	2905	3194	4221	3600
USA	2068	1228	1159	1069	970
Japan	5184	5786	6035	5710	6020
OECD	22408	22376	22148	19222	17569
World	31813	36502	40542	40328	41490

Source: Table 5.3, Transport Annual Statistics, 1970-1990, Office for Official Publications of the European Communities (1993), p.134.

However, offset against these effects are:

1. Reduced maritime employment opportunities.
2. Tarnished public perceptions of shipping.
3. Associated impacts on manpower recruitment levels, in a range of economies.

Recently, other work (Li and Wonham, 1999) has argued that some of the more gloomy predictions of a future shortage of qualified seafarers be treated with at least a measure of caution. In particular useful distinctions need to be drawn as to what constitutes, for example:

1. A qualified seaman. Chinese seafarers, comprising the largest single national pool, who might be considered to be unfamiliar with the language, culture or operating styles of some non-Chinese fleets, still

constitute a highly active world resource, even if not considered to be 'qualified' to work under other flags.

2. An employed seaman. Shortcomings in UK national statistics collected by the Chamber of Shipping since 1990 for example, exclude seamen employed by non-members, or may include other seamen on the ships of members not flying the UK flag.
3. Assumptions of the widespread applicability of average growth rates. Due to variations in national conditions, these may be flawed, as may any assumptions that the fleet growth can be deterministically linked with the demand for seamen.

The implications of these distinctions are that the human resource availability debate should shift from one solely based on the quantity of seamen available, to focus more on the issues of their quality. If this can be raised to acceptable levels, then taken worldwide, sufficient human resources are available and if harnessed, could meet overall operating requirements. In the light of these recent trends, it is useful to consider some of the policies that the shipping and logistics industries and their constituent organisations might seek to adopt in these circumstances.

Some Industrial and Organisational Responses to these Trends

In the light of increasing pressures on the skilled human resource base available to shipping and other international businesses, theoretically, a range of industrial responses could be adopted. Responses considered here include 'factor substitution' in the supply chain where the ratios of capital and labour inputs may be varied, better strategic planning or increased productivity resulting from a better educated labour force.

Supply Chain Re-engineering

A strategic desire to improve customer service levels through 'time compression' in the supply chain has probably provided the dominant motivation of industries or organisations seeking to re-engineer their supply chains. Following in their wake, holistic reviews of the ways in which supply chains are managed have been stimulated (Bumstead, 1998, p.160). In the face of shortages of skilled manpower, and given that commodity or bulk freight movements by sea usually form an indispensable element in many international supply chains, an obvious corporate response might be to employ new technologies to substitute for manpower where possible. Although such polices are potentially attractive, their immediate worth is limited as Obando Rojas et al (1999, p.41) noted where:

> Technological advances in shipping involving an increased use of automation have undoubtedly worked toward this end [supply chain re-engineering] by reducing manning requirements. However, unless the unmanned ship becomes the norm, technology alone cannot solve manpower shortages.

This policy is at best viewed as one element of a very long term solution, which must also include other components conceived within a holistic review of the entire supply chain.

Strategic Planning

A range of approaches to strategic management, pertinent to decision makers in maritime businesses have been reviewed elsewhere (e.g. Ferch and Roe, 1998, p.6-35). However in terms of the deck officer manpower problem, one particular strategic model which analysed the structure and polices of this particular system has proved useful. In adopting a systems dynamics approach, Obando Rojas et al (1999, p.42) explicitly highlighted short, medium and long term perspectives of a complex system, in which flows of information, people and capital were modelled. A key element in their analysis of the UK supply chain identified the 'recruitment subsystem', where factors contributing to the decline in recruitment included, 'largely unresearched and unaddressed ...the poor maritime awareness and the lack of a clear professional career structure in the merchant marine' (p.49). The authors then proceeded to raise issues relating to the narrow range of subjects studied on professional courses, observing a requirement to include studies of a broader range of academic disciplines including business and law. Also mooted were funding issues, where an increasing responsibility is being placed on individual students having to find their own funding for higher level education, rather than relying on traditional corporate or governmental sources.

Educational Provision

Obando Rojas et al (1999) also noted that recruitment and retention formed the key elements in balancing the supplies of and demand for seagoing officers. Perceptions of seagoing careers and their structures and funding opportunities were likely to influence the former, whilst a range of psychological, occupational, economic and technological issues influenced the latter. The ways in which relevant postgraduate study programmes are perceived and provided may potentially impact on a several of these issues. In terms of recruitment into the maritime industries, it is likely that many applicants to postgraduate courses would regard a pertinent postgraduate vocational qualification as a preparation for a shore based career. If so, such courses could potentially furnish at least a partial substitute manpower source for the diminishing supplies of experienced seafarers who have traditionally filled these roles. Although the provision of high-level academic courses is unlikely to influence the supplies of seagoing cadets, more interested in entry level qualifications, they might also influence mid career retention rates, with obvious implications for the motivation and organisational commitment of individuals who partake in them. Not only have individuals who embark on lifelong learning programmes been observed to be more flexible and employable, but also better motivated to achieve their long term career goals (Otala, 1994, p.207). In this sense, an understanding of the perceptions of students embarking on vocational postgraduate courses could prove kernel to influencing mid career retention rates, and recruitment into at least the shore based, and possibly other maritime industries.

Individual Responses to Trends: The Study of Maritime Subjects

In order to prepare themselves to face this competitive environment, one response by individuals might be to upgrade their educational qualifications. However, before reviewing research relating to the reasons for choosing to study maritime subjects at university, some allusion to the traditional attractions of a seafaring career is appropriate. Recent attempts to integrate theories of academic and career choice in individuals (Lent et al, 1994) also hint that such considerations may assist in understanding why students choose particular university courses.

Opportunities for travel were known to represent the main traditional stereotyped attraction of a seafaring career (e.g. Hope, 1980; Frickle, 1974), but were rarely annoyed by practitioners who in reality spent very little time ashore whilst abroad. As the numbers of British seafarers declined, so also did the glamour associated with their lifestyle, being replaced by more realistic understanding of the pay levels involved, and their hum-drum existence (Hibbs, 1988, p.66). Premature termination of increasing numbers of careers in the merchant navy prompted research (Board of

Trade, 1970) into why this was so, revealing that many deck officers had entered their careers with limited long term career commitment. Coupled with the unrelenting demands of an ever changing workplace location often far removed from their homes, many seaman were tempted ashore by marriage at relatively early stages in their working lives, and proved unwilling recruiting sergeants for their replacements. Often, only those youngsters drawn from families with a long seafaring tradition possessed realistic expectations of life at sea.

Studies of the perceptions of the social and economic status of merchant navy officers (Boulter, 1990) have revealed differences between the views of teenagers living in coastal areas and those living in inland areas. In particular, respondents in coastal areas have been observed to view the pecuniary status of seafarers as being more favourable than those in inland areas, although both groups were also found to associate travel, responsibility and adventure with this occupation. In the same study, those respondents with relatives or friends at sea rated the status of seafarers more highly, and were themselves more attracted to a life at sea. Media images, often negative, such as those associated with disasters, were found to influence perceptions of the attractiveness or otherwise, of careers at sea.

A few studies have been conducted investigating the reasons why undergraduates seek to embark on higher education courses generally (Dearing, 1997) or the attractions of particular vocational specialisms including marine studies (Dinwoodie and Heijveld, 1997), maritime business (Dinwoodie, in press) and transport (Dinwoodie, 1996). Such research has revealed a tension between the vocational motivations for undertaking advanced study, and academic enjoyment, with the former becoming more prominent as graduation approaches. Although freshmen have been attracted initially by 'love of the sea' on matriculation, growing concerns for their future basic needs relating to pay, status, responsibility and good prospects appear to predominate for graduates (Dinwoodie, in press). There is evidence that other attributes of such employment, including travel, a dynamic workplace, and separation between home and the workplace, are known to, and generally accepted by students, as inherent characteristics of their chosen workplace.

In responding to similar problems, logisticians devised a series of measures to stimulate interest in young people (Mills et al, 1999). These included a recruitment video, seminars for careers advisors, school packs, business games for undergraduates and specialist careers guides, although their longer term effectiveness is not yet known.

When recruiting new entrants to the industry, employers may consider industrial 'experience' to be of greater value than purely academic skills in candidates (Evangelista and Morvillo, 1998, p.91). Weight is given to interpersonal skills involving teamwork and communication, and personal traits including patience, initiative, enthusiasm and responsibility. Even if students succeed in attaining such skills whilst in the classroom, it is still not guaranteed that graduates will realise their vocational expectations during their working lives (Tsakos, 1996, p.27), if they possess unrealistic perceptions of their chosen careers.

Many highly qualified and experienced seamen, should they seek to change careers or go ashore, are likely to find that their hard earned seagoing skills and competencies are held in lowly regard by other potential employers. On the one hand, this phenomenon represents a dual assault on the supply of manpower in the shipping industry, where not only could it become almost impossible to recruit suitably skilled non-seafaring professionals directly into shipping employment, but also, once trained, the competencies of seafarers have limited exchange value. It becomes harder to attract young people into seagoing occupations when they may fear that the experience and expertise which they painstakingly acquire is likely to be redundant by mid career. Wastage in shipping was long since conceived of as a

> one-way flow, i.e. when our trained officers and ratings leave the sea they are not only lost to the shipping industry but they cannot be replaced by men trained in other industries (Moreby, 1968, p.6).

On the other hand, individual seafarers must also face the prospect of extensive retraining and likely related loss of income, should they wish to go ashore.

> Some highly qualified officers do find employment ashore, but this is usually in the maritime or kindred industries where their education and training is directly applicable. These are a minority and do not undermine the point that the industry has a highly skilled and experienced work force who lack general transferable skills (McConville, 1999, p.207).

The issue of a lack of emphasis on transferable skills development in traditional maritime courses might appear, *prima facie*, to be a prime concern for academics, but it has much wider ramifications. Potentially, the skills set that an individual who embarks on such courses might expect to acquire, and hence their market value in the workforce forthwith, might depend on it. Equally, it can also influence the perceived attractiveness of courses in international shipping and logistics to potential applicants, and hence future industrial recruitment levels. These issues are important concerns in this volume.

Aims and Objectives of the Book

The aim of this work is to investigate and report on the reminiscences of individuals, who possess managerial experience or aspirations, into how they decided to embark on advanced study in shipping and logistics. The work employs a variety of research techniques, in order to understand more fully how high calibre recruits are attracted to these courses. In this way an improved understanding will be gained into:

1. Characteristic attractions of courses in shipping and logistics to potential employees and managers.
2. The particular problems of preparing postgraduate students to embark on international careers.
3. Devising bespoke research instruments and methodologies appropriate to this educational group.
4. Methods by which the management of vocational marketing and recruitment campaigns might be improved.

Methodology and Outline

In order to achieve these objectives, work is presented which:

1. Investigates the perceptions of how students came to choose particular courses.
2. Develops and evaluates instruments with which to assess how students perceived that they had arrived at their decision to study at a particular university.
3. Analyses factors that influenced the perceptions of groups of students.
4. Recommends ways to promote recruitment into advanced vocational courses, maintaining high entry standards.

In Chapter 2, the environments within which universities undertake to provide advanced courses of study are considered often in an international setting. Both the pressures placed on universities to recruit international students and influences on the decisions of individuals to undertake study are relevant. Literature relating to preparing future employees for international careers is also discussed.

How do student perceptions of employment roles in shipping and logistics evolve? If such perceptions are not understood, it is impossible to build on and develop students' existing knowledge, to review any characteristic attractions of particular occupations or to define their role in attracting students into vocational courses. In Chapter 3 examples of how such perceptions evolve and some pertinent research techniques are presented.

Qualitative empirical work based on focus groups exploring individual decisions to study international shipping and logistics at postgraduate level at Plymouth is presented in Chapter 4 and quantitative results of applying a bespoke instrument to analyse the study decision are discussed in Chapter 5. In Chapter 3, there was clear evidence of an evolving understanding of individual perceptions of employment roles. Based on this finding, an attempt is made to map and compare individual perceptions of the decision to study international shipping and logistics using an explicit cognitive mapping approach, reported in Chapter 6. A worked example of structuring a decision as to where to attend a short course is also presented. Further empirical and theoretical explanations of these observations, and attempts to make comparisons between particular subgroups and explain the factors influencing them, are presented in Chapter 7.

In Chapter 8, the discussion is broadened to compare the perceptions of undergraduates and postgraduates, at UK national level, of their reasons for undertaking advanced study in maritime business and logistics. The discussion is further extended in Chapter 9, to consider the need for industrial practitioners to engage in continuing professional development. As a case study, a survey of the attitudes of practitioners towards port state control in the UK is reported, along with an outline methodology for devising a strategy to better inform and educate participants. However, this methodology could be more extensively applied. Finally, in Chapter 10, some of the implications for employers and operators, those considering study, human resources managers, providers and marketers of courses and academic researchers are outlined.

2 International Students, Universities and Recruitment

Introduction

In this chapter we first consider the environments within which universities undertake to provide advanced vocational courses designed for an international clientele, and within which prospective students seek to upgrade their education and competencies to meet business requirements. It is important to be aware of the pressures, financial and otherwise, which have stimulated UK universities to recruit increasingly large numbers of international students. However, unless these pressures are handled responsibly, adverse reports of inadequate student support systems may filter back from alumni to prospective applicants, diminishing a potentially lucrative source of university income. Whilst a desire for vocational advancement may motivate some individuals to undertake advanced study at university, perceiving it as raising their future employability, all applicants must still balance their academic interests against more basic motivations of increased expected rewards for doing so. In the case of individuals seeking to engage in future international careers, there is evidence to suggest that study abroad can form a useful foundation for establishing the personal qualities demanded by such a lifestyle.

At a corporate level, the forces of globalisation present organisations with a need to recruit, advance and retain employees capable of working within a range of cultures and business environments. In human resource management terms, these individuals must not only be empowered to fulfil organisational objectives, but as the bitter experience of organisational learning has indicated, their personal development needs must also be met if they are to be retained. Postgraduate study may present one way of meeting these development needs, and the argument changes tack by discussing ways in which postgraduate students may be prepared for their subsequent international careers. The range of competencies and qualities viewed as important by employers of individuals engaged on international assignments, are reviewed briefly. Next, models of business thinking appropriate to future international careers and a review to an existing model of strategic international career management are outlined. Finally, policies designed to assist postgraduate students to develop requisite skills whilst in the classroom are discussed. Throughout this chapter, these issues are discussed broadly, before being considered within the specific context of international shipping and logistics in later chapters.

Pressures on Universities to Recruit International Students

University managers must recruit overseas students responsibly, trading a potentially lucrative source of revenue against a resource intensive liability (Williams, 1992, ch.6), where failure to provide adequately for them may generate adverse reports which filter through to funding councils or potential student recruits. About 20 per cent of the UK's growing postgraduate taught course students, some 50,000 in total, are from overseas (HESA, 1997, pp.12,16; Parry, 1997, p.11), and the implications for the recruitment processes on which this delicate balance depends are crucial to university managers. In order to inform decisions, it is necessary to investigate and understand the decisions of individual students to embark on different types of postgraduate taught courses. In the current work, the decisions of postgraduates to study international shipping and logistics at Plymouth are the main focus.

Explanations of the growing interest in internationalisation in UK institutions of higher education range from the innate to the pragmatic. If a university is conceived literally (universitas – 'the whole world'), any claims to mere regional or national hegemony in learning become a contradiction which deny its universal eminence. In adopting such a view, internationalisation becomes a

> defining feature of all universities, encompassing universal organisational change, curriculum innovation, staff development and student mobility, for the purposes of achieving excellence in teaching and research
>
> (Rudzki, 1995, p.423).

From this stance, Rudzki developed a strategic management model of the process, which articulated each of these elements. By contrast, Harris' (1995) more pragmatic view was based on increasing pressures on UK universities to seek funding outwith the funding councils through setting ever more competitive fees in the market for overseas students. Such institutions, possessing only finite resources with which to support international students ever more prone to language problems (Cownie and Addison, 1996), socially isolated and remote from home (Johnston, 1995), face increasing demands on their support facilities. Whatever its causes, few UK universities can afford to shun the ramifications of internationalisation, and yet they are not widely researched (Harris, 1995).

Increased funding pressures on UK universities in the 1980s promulgated a need to seek alternative sources of finance, including marketing and recruiting initiatives directed at overseas students (Williams, 1992). Historically, motivation for Government support of overseas students has ranged from enlightened self interest which aims to cream off talented researchers, or win friends by ensuring that sufficient future leaders have studied in the UK, to fostering political relations in the national interest and the altruism of overseas development programmes. However, following the shift from indiscriminate subsidy for overseas students in the UK before 1980 to full cost fees and later selective subsidy, specialist international offices emerged in many British universities. These bodies offered professional marketing of courses generating high revenue to cost ratios, reflecting the increased pressure to apply business logic in this context. Concomitant with this approach, it also became necessary for universities to review the nature of the product they supplied, and raise the quality of service offered to their customers. Although universities were primarily concerned with pedagogic issues such as course content, language support, modes of delivery and assessment, and class size which underlie the former, these factors may have less direct influence on the decisions of potential students to select particular courses and institutions than the latter. Arguably, more immediate issues such as the perceived living conditions including accommodation, welfare provision, financial support and immigration problems, may be more important to some potential recruits (Woodhall, 1989).

Responsible Recruitment

'Responsible recruitment' by universities implies higher costs of 'customer care' which may in turn eat away at the profit potential of excessive numbers of overseas students. Mature students for example, may wish to bring families with them who require special accommodation, with specialist counselling and support services, which must be available throughout their stay in the UK, and dedicated social or meetings facilities may be needed for particular ethnic groups. Unless high quality specialist student support is provided for each of the groups recruited, poor oral reports about institutions which filter through to potential recruits are likely to expose and rapidly undermine not only the excesses of individual institutions driven by myopic greed, but also the host nation. Typically, a life cycle approach to each individual overseas student experience

> ...must be catalogued from the point of selection ...to the return home ...where it is normally hoped that as loyal and appreciative alumni they will serve subsequently as volunteer recruiting sergeants (Harris, 1995, p.83).

When being recruited initially, it is doubtful whether individual overseas students are aware of many of the trials and tribulations that are likely to await them while studying in the UK. A growing body of evidence (Jochems et al, 1996) suggests that even though success rates for overseas students are similar to those for native students, they may need more time to pass their examinations. In particular, their grades in examinations may be lower, with multiple attempts required in order to achieve success, and a greater probability of having to postpone examinations. Solutions requiring personal mentors and more resource intensive teaching for newcomers to courses may conflict with the institution's original intention of raising its resource base by recruiting them.

Isolation can be a major problem for overseas students, particularly among females (Conrad and Phillips, 1995). At doctoral level, a lack of specialist counselling facilities and an impersonal set of social contact opportunities may make them especially prone to emotional problems, resulting in lower thesis completion rates. Solutions involving novel modes of communication, including more informal induction programmes and research seminars, computer conferencing, and more supportive forms of interaction by supervisors, may all place increasing demands on resources available within institutions. The induction process was found to be critical (Johnston, 1995) in building a sense of belonging, especially where many students may be part-time, mature or female. Courses may involve little opportunity for discussion, or indeed contact of any kind with fellow students, demanding long periods of independent study. Newsletters, information booklets, seminar and research conferences and meeting rooms may help to relieve some of the sense of isolation, but unless students felt a part of the department's research culture, they were unlikely to thrive.

Problems and Benefits of Undergraduate Exchange Programmes

The experiences of undergraduates undertaking exchange study programmes in Europe are relatively well researched (Burn, Cerych and Smith, 1990; Teichler and Carlson, 1990; Teichler and Maiworm, 1994), and offer useful retrospective insights into how this particular group approached the decision to study abroad. In surveys of ERASMUS students, all but five per cent reported that they had undergone specific courses preparing them for their exchanges. These included training in languages, practical living abroad, cultural issues and even unfamiliar academic systems (Teichler and Maiworm, 1994, p.7), with those institutions which offered high degrees of assistance gaining the highest ratings from students. Despite these preparations, the most common problems cited by returning exchange students related to living and organising the conditions of study in the host country, including concerns over finances and accommodation, social contact and foreign language problems (Teichler and Carlson, 1990). However, a larger proportion of graduates who had undergone ERASMUS exchanges, some 50 per cent, subsequently undertook postgraduate study, compared with only 30 per cent of all EU graduates. This may indicate either that undergraduate study overseas is more likely to encourage general interest in further study, or higher levels of educational motivation in this group. When classified by academic discipline, the proportions of exchange students undertaking further study varied from over 70 per cent in more academic subjects such as geography and social sciences, to only 34 per cent in business studies, more akin to logistics and shipping courses.

Experience of living in a foreign country was considered to be an important dimension of the undergraduate exchange process, by students and managers alike. When programme directors were asked to rate the expected impacts of undergraduate study abroad programmes, they cited improved communication with foreigners, individual development, language skills, wider awareness of the subject area and teaching methods, ahead of enhanced career prospects (Burn, Cerych and Smith, 1990). Similarly, in terms of the profile of a typical exchange student, two-thirds of European participants had already spent at least a month abroad after the age of 15 prior to exchange, mainly as tourists. Significantly, the parents of many exchange students had completed courses of higher education, including 55 per cent of fathers and 40 per cent of mothers, but with rates in the US as a home country more than double those in the UK. Overall, the reasons why European students had chosen their home institution included, in descending order of preference, subject area interest, vocational interest, possibilities of study abroad, career prospects, personal strength in the subject area, or the institution had been recommended to them. For business students, the general desire to study abroad rated ahead of strictly academic issues. The prospect of study programmes abroad was the most important factor, followed by the nature of the programme offered, particular institutional strengths and prestige, its location and proximity to work or family, non admission to another institution or chance. Motives for European business students wanting to study abroad included language skills, desire to live abroad, improved career prospects, understanding of the host country, desire to travel and less importantly, reasons such as friends who were also going, and a break from their usual surroundings. Eighty five per cent of European students claimed that the decision to undertake such study was their own.

Increasingly, the undergraduate study experience per se is less likely to prove sufficient to meet the ongoing educational requirements of many individuals. Lifelong learning may become a necessity in the face of changing technology and information systems, as dictated either by work in changing fields such as shipping and logistics, or in society, by helping people to meet their changing social, emotional or aesthetic needs (Knapper and Croppley, 1991). Willingness to engage in learning is a fundamental prerequisite to lifelong learning, and individuals must possess the right attitudes, values, self-image and study skills, in order to proceed further. An extreme view might argue that the willingness of individuals to undertake further study at successively higher levels is one reflection of the effectiveness of their earlier preparation in both helping them to want to learn, and learning how to learn, implying a need to develop such skills in courses at all levels. Students are only likely to wish to progress to postgraduate courses, if undergraduate programmes have succeeded in developing their ability and willingness to do so. However, competing demands on lecturers' time may limit their potential as a source of information for aspiring undergraduates at critical career decision points, where such interaction, however important to individual students, is not highly regarded by faculty (Boyer, 1990, p.1).

Mature students face particular problems when undertaking study, and may be deficient in particular types of skills (Richardson, 1995). In a comparative study of mature and non-mature undergraduate students, the former performed at least as well as the latter overall, but their approach to study differed. In particular, mature students were less reliant on memory skills than younger students, but were more likely to display a deeper approach to learning, concentrating more on meaning in their studies. As such, unless teaching methods give mature students more scope for active questioning, allowing ideas between parts of a course to be more interrelated, with more scope for divergent thinking and relating evidence to conclusions, they are more likely to experience learning problems than younger students.

Funding may present yet another obstacle to completing a graduate course (St John and Andrieu, 1995) in that increases in tuition fees correlate positively with the time taken for an average student to complete a degree, and negatively with the proportions who complete their degrees. In order to increase completion rates, a mixed funding package including loans, grants and assistantships, was found to be more likely to succeed, although other forms of funding have been observed in practice (Tight, 1992).

Other literature is pertinent to the reasons why, or processes whereby, overseas students may seek to undertake postgraduate study. In the former context, Ainsworth and Morley (1995) found, ex post, that graduates from a Master of Business Administration course viewed their studies as having enhanced their career prospects, regarding them as relevant, and offering them increased knowledge and behavioural changes, with few differences attributable to gender or employment sector. The recruitment policies and methods which universities may adopt to woo overseas students have been catalogued (Woodhall, 1989, p.146) and include such activities as advertising, lecturers distributing literature abroad during visits, promotional visits, alumni associations, recruiting agents and bilateral links with institutions in the UK and overseas. However, the viewpoints of applicants, prior to undertaking study, were not their prime point of reference.

Why Do Individuals Decide to Undertake Advanced Study?

Traditionally, many companies or even the state have initiated and funded training schemes that have benefited individual employees or citizens, in addition to their sponsors. However, individuals are increasingly being requested to take on greater responsibility for defining, managing and funding their own training needs. In situations where the survival of a business may depend on the ability of its human resources to adapt rapidly to external changes, policies that accelerate this process may in turn stimulate the business. Observations which show that individuals who take on more responsibility for managing their own training needs become less resistant to change (Baldwin, Magjuka and Loher, 1991), imply a need to extend the trend.

Some recent academic developments have also attempted to redefine the role of the individual, such as the attempt to build a single unified social cognitive theory of career and academic interest, choice and performance (Lent et al, 1994). This theory may help to explain early career and academic choices, but may be less well suited to modelling mid career changes made by experienced postgraduate students (Panayides and Dinwoodie, 1999). However, the view that career interests and goals, previously thought to be closely linked, are mediated by a range of barriers to achievement is a useful one. Such barriers may include ethnic or gender issues for example. Studies of such perceived barriers in groups potentially disadvantaged in their career decisions, found merely that more subjects were likely to agree that certain barriers would affect them, even though the majority did not agree (McWhirter, 1997). Although systematic study of gender and ethnicity barriers would be interesting, the prime current concern is to trace evolving perceptions of a single study decision, rather a general career decision.

A further compelling reason why individuals undertake study, is peer pressure or a professional requirement to do so. By 1999 some 25 per cent of UK logistics managers held first degrees, with a further eight per cent holding second degrees (Rogoff, 1999, p.41). Also, although not a condition of a right to practise, in excess of 20,000 logistics managers in the UK are members of the Institution of Logistics and Transportation. Obtaining a relevant degree is one of the ways of satisfying the academic requirements for membership of relevant professional bodies.

The Role of Career Exploration Behaviours

Uncertainty of employment in the workplace, has been brought about by a shift to a global scale of corporate operations, coupled with social and technical changes which demand that the skills of employees must match them (Goldstein and Gilliam, 1990). Individuals who seek to remain in employment face a lifelong process of having to find funding to acquire new skills with which to equip themselves for the changing workplace (Knapper and Croppley, 1991). However, this process provides only one possible context within which to view the desire of individuals to undertake postgraduate study. It is important that university managers and admissions tutors striving to design and market attractive postgraduate courses should also seek to understand the broader processes pertaining to how and why individuals seek to embark on such courses.

Earlier thinking which attempted to identify and classify the vocational behaviour of individuals into distinct career stages has recently had to be reviewed (Super, 1990). In particular, as technological and organisational changes create redundancies or redefine individual circumstances, employees may need to revisit earlier career behaviours at several points in their working lives as their careers are 'recycled'. It may be appropriate to view a decision to undertake postgraduate study, at any age, as a career exploration process. Alternatively, it may be viewed within the context of a lifelong process of matching the changing needs of the world of work to changing individual aspirations (Hall, 1992). It is expected that such student interest in finding out about courses should diminish after the study decision has been made.

Towards graduation, students might logically be expected to think ahead towards the next stage of their careers, with growing interest in establishing themselves (Smart and Peterson, 1997, p.371). Also, after a decision to study has been made, the stated importance of the attractions of study in a particular subject, country or institution will probably exceed those prior to it, as the decision is 'justified'. In the same vein, the perceived importance of deterrents and barriers to embarking on study generally, or courses in particular, are likely to be greater before the decision is made.

It is now necessary to consider the way in which postgraduate education might begin to prepare aspiring managers for their future careers in international business. In particular, the discussion will begin by considering some of the competencies that employers are likely to demand of employees likely to undertake such assignments in the course of their work with them. A strategic model of careers for international management is then discussed, along with a range of modifications that have also been proposed. Finally, some of the implications of these approaches for teaching in the postgraduate classroom are considered.

Employee Competencies Required for International Assignments

According to one view (Bartlett and Ghoshal, 1992), global organisations may employ a range of managers involved in international operations, but the phenomenon of a single universal international manager is non existent. Functional specialists, often engaged in technical roles, may operate in any location, whilst national managers operate mainly within one country and one culture. At a senior strategic level, only a few international managers may be responsible for ensuring that best practices within the organisation are disseminated globally enabling all other staff to operate within clear hierarchies. Their paucity may reflect a shortage of competent individuals available, or selection biases including gender (Lineham, 2000, p.129).

An observed failure of one-third of secondments, rather than being due to culture shock, may result from a failure to ensure that seconded executives possess relevant competencies (May, 1997). Requisite business skills include a need to think long term, move between functions, work in multicultural teams, take an overview and 'think locally, but act globally'. Desirable personal competencies, aside from the obvious such as language and communication skills, resilience and an ability to cope with stress, include cultural awareness, toleration of ambiguity and empathy with local values. Postgraduate classrooms can provide an ideal forcing house for acquiring such competencies.

Cultural sensitivity and empathy are important aspects of expatriate competence. However, although these concepts are difficult to teach, through the use of cross-cultural training, international managers can learn how to adapt to cultural differences (Webb, 1999).

Effective communication between two parties involves a mutual understanding of the semantics of the ideas and concepts which are exchanged, and not merely the successful physical transmission and receipt of syntax and signals. In order to participate in such an exchange, would-be players must have acquired a knowledge of the rules governing how acceptable exchanges are made, which may in turn be dependant on having gained membership of a shared system within which common social exchanges are performed. An ability to function within this cultural system implies a cognitive awareness of its constituent elements, exhibiting appropriate behavioural patterns, displaying and interpreting correctly the symbolic representations within which its information exchanges occur, and an assimilation of its value systems. Through 'telling stories' language may be used to establish whether a social relationship is strong enough to share exchanges built on constructs with only local meaning (Gold, 1998), and by implication, to exclude the uninitiated.

Individuals who seek to do business within a culture must genuinely want to understand and communicate within it, and be sufficiently flexible to adapt to its values. Cross-cultural training programmes that aim to impart knowledge to employees about the new cultures within which they will be interacting may also increase participants' confidence in forming new relationships and increase their awareness and adaptability skills.

On a postgraduate course, excessive concentration on regionally specific skills may be inappropriate, except where a professional culture, such as shipping, shares a common international language, such as English. On a course where participants' attendance may equally be dependant on funding by themselves, or their companies or national governments, or varying combinations of all three, the concept of attempting to interact within a single organisational culture is inappropriate. However, a willingness and ability to tolerate, appreciate and even celebrate the mixed messages which may emanate from other course members is essential, and surely an excellent forum within which to participate in cross-cultural training. An ability to adjust into a multinational classroom surely presents evidence of at least the potential to thrive on an international corporate assignment, which may outweigh the role of technical considerations.

> Ninety per cent of the time, businesses select employees for overseas assignments on the basis of their technical expertise, not on their cross-cultural fluency. Rarely are traits such as cultural sensitivity, interpersonal skills, adaptability and flexibility taken into consideration (Webb, 1996, p.2).

The total costs to a business of employing staff overseas are several times those of employing them domestically, and finding the 'right' person is vital, in order to recoup these costs and ensure that potential business is not lost. Whilst it is impossible to define a single 'right' type of person for all overseas assignments, a generic set of personal characteristics might include those shown below (Webb, 1996). Employees must be able inter alia to:

1. Adapt to changes, both anticipated and otherwise.
2. Cope with the stresses of adapting to new cultures.
3. Possess a sense of humour.
4. Make, build and maintain a wide range of business and social relationships.
5. Appreciate and respect behaviours and attitudes contrary to their own.
6. Communicate with and interact with people from other cultures.
7. Work independently.
8. Communicate a non-authoritarian management style.

For non-British nationals, be they European or non-European, one of the attractions of undertaking an international postgraduate course of study in the UK, is to learn to develop these personality traits in a non-threatening environment. In addition, study in the UK presents a unique opportunity to enhance linguistic skills in English, the language of international shipping, and to acquire grounding in its business and cultural environment. Students who can learn to adapt to this relatively non-threatening environment are also more likely to be able to convince potential future employers of their suitability for careers within an international business environment. Such attractions are likely to appeal to all types of student, although there may be gender biases in the attractiveness to graduates of working abroad (Counsell, 1996), with higher proportions of females expressing such interest. However, many of their career aspirations may not be realised in practise, with significantly more males than females actually gaining expatriate employment (Lineham, 2000). It is now time, to review some of the models within which business thinking appropriate to an international career have been conceived.

Models of Business Thinking Appropriate to an International Career

Is any particular model of business thinking appropriate to a career in international business? Cross-cultural training for future employees likely to be involved in international assignments will aim to develop an individual's ability to communicate and empathise within the context of a range of national environments. In preparing them for international careers, students may need to be taught to question the historical origin of the cultural values that may taint career management issues. As one example, ideals of high technical competence and good social skills associated with German organisations may be suspect, where surveys revealed that employees expressed a need for better people skills (Hansen and Willcox, 1997), with training in such skills restricted to managers. Similarly, the need to empower younger employees with more information, which they were found to value more than financial rewards, conflicts with the needs of older employees who merely tended to hoard it, fearing job insecurity.

At the level of analysing individual behaviours, career transition theory (Nicholson and West, 1988) appealed to notions of recursion, interdependence and discontinuity. Although constantly changing, or recursive, an interdependence occurs between career stages where the events of one stage influence those of the next stage, with discontinuities between stages defining distinct experiences and problems at certain stages. Four modes of role adjustment are possible, including replication, absorption, determination and exploration. In the replication mode, existing attitudes and behaviours are transferred into the new role settings, whilst in absorption mode, the onus is on the individual to learn the new skills and behaviours associated with the new role. In determination mode, the role rather than the individual is adjusted to the new situation, and in the exploration mode, both the role and the individual adjust to the new circumstances, presenting the potential for changes in both corporate operations as well as individual attitudes and behaviour.

At a more strategic level, Adler and Ghadar's (1990) model proposed a matching of corporate international goals and organisational human resources management practice, whereby the environmental influence of, for example, international business, promulgates internal changes in the organisation's career management system. However, the details of cultural adjustment in the early stages of expansion overseas, or readjustment in recessional phases, or detailed changes within particular phases, are not defined.

Fish and Wood's (1997) model of strategic international career management focused on both the organisational and individual elements of career management, by fusing Nicholson and West's (1988) approach with Adler and Ghadar's (1990). Justified by a concern to ensure that career patterns are not out of kilter with cultural adjustments and changing business strategies, they proposed a framework within which to minimise the probability of failure due to maladjusted individuals or organisations.

Adjustment processes were defined for both the organisational and individual contexts. Within the organisational context, stages included:

1. Preparation (selection, training, job design);
2. Encounter (induction, socialisation);
3. Adjustment (appraisal, compensation, training); and
4. Stabilisation (career advancement).

In the preparation phase, transitional career elements are apparent as managers acquire the skills and awareness needed for international assignments. As managers prepare to go overseas, adjust to new conditions and prepare to return home, recursion is apparent. Where managers have been well prepared for an assignment, interdependence with the adjustment process, should be reflected in the latter. Discontinuities are apparent where the role of a logistics manager may differ as the range of technical possibilities for managing physical or information flows vary between developed and less developed economies, for example.

In ensuring that individuals are developed responsibly, their background, competencies, and motivations must suit the role proposed for them. They must be committed to the organisation and personal issues must have been addressed. Along with self-efficacy and previous overseas experience, these issues define the preparation responsibilities. Induction and personal development define the encounter and adjustment stages, while preparing for the next move or commitment to stay define the stabilisation phase.

In ensuring that individuals are competent for their new assignments, and later assisted to adjust to it, transitional elements are apparent. Where individuals are prepared for repatriation, recursion is apparent, and when, where individuals accept assignments as part of a planned career development process and their adjustment is eased, interdependence is apparent. If an individual becomes overly attached to one assignment, discontinuities between the needs of the individual and the organisation may become apparent.

Specialist education has a role to play in all four of these stages. Pertinent postgraduate courses may be perceived as furnishing individuals with the cognitive skills sets which their international careers will demand, and if overtly designed so to do, can also develop the affective and attitudinal dimensions which new entrants will require.

1. Where individuals have successfully completed courses of study overseas in multinational groups, their experience of successful transitions should be apparent, in raising both their individual self-efficacy, and providing organisations with evidence of their ability to adjust to new cultures.
2. By invoking the concept of interdependence, individuals who have been encouraged to contemplate realistically their future career moves before completing a course of study, should be better able to adjust to new assignments.
3. The concept of recursion implies a need to debrief students prior to leaving courses, helping them to reflect on how they will approach the demands of preparation and encounter phases in their subsequent assignments.
4. Discontinuities may be exaggerated, for example, where overseas students hold unrealistic expectations of obtaining permanent employment in the country in which they studied.

Implications for Preparing Postgraduates for International Careers

Stumpf (1998) identified a range of skills associated with successful course design and delivery in corporate universities, engaged in the business of producing the quality leaders who will enable future corporate objectives to be realised. Some of the skills demanded of a successful 'whole manager' with a global perspective, such as making presentations, negotiating and teamwork are readily apparent. However, other determinants of success might be dependent on opportunities to develop such skills as problem finding and diagnosis, creative problem solving, ethical decision-making and career management. Regarding the latter, many employees have been observed to be confident with the self-review and exploration process, but less so with actually making choices (Stevens, 1996), requiring counselling support. This implies a need to tutor postgraduate students in these skills, to prepare them for a volatile employment market.

In an era of lifelong learning, it is acknowledged that managers can expect to return regularly to the classroom (Fulmer and Gibbs, 1998). They will need to be continuously open to new ideas, and courses of study must prepare them to continue to face the future, as flexibly as possible. Training in shaping attitudes to jobs may be more useful to them than job specific skills, particularly for aspiring managers of international operations.

What makes an effective management education programme? Training must be intensely practical, and angled towards an action plan for changing behaviour in the workplace. Active learning is most effective, although case studies, discussion and simulation may also be useful (Longenecker et al, 1998). At the end of the day, attitude is again critical, with a need to ensure that all parties involved take the whole exercise seriously. Unless this is so, a bad programme can waste a lot of valuable resources, creating mistrust and negative outcomes.

Summary

Postgraduate vocational taught courses constitute an increasingly voluminous and competitive market place. However, especially so in International Shipping and Logistics, this has not been extensively researched from the perspective of the aspiring student. From a university's perspective, gone are the days when course profitability was a simple function of the number of recruits. These were long since replaced not only by a need for responsible recruitment which ensures adequate student welfare and support provision, but also a discerning clientele who shun courses perceived as shoddy or irrelevant to their needs.

When deciding to apply for vocational university courses, the lure of enhanced employment prospects on graduation might, prima facie, be expected to be significant to aspiring students. However, before embarking on an empirical assessment of the influences of employment and other considerations on the decision to study international shipping and logistics at postgraduate level, it is first instructive to explore how perceptions of similar employment roles evolve. The prime concern is to establish and test an appropriate research methodology, which demands an attempt to understand the processes operating which determine how student perceptions of employment roles in the transport field evolve. The effort offers the prospect of both furnishing an armoury of useful research techniques and evidence of evolutionary sequences of thought processes, which will assist in addressing the main task at hand. In this discussion, the technique of quantitative content analysis is introduced, which will also be used to establish and analyse the range of issues and concerns felt to be significant by postgraduate students.

3 Evolving Perceptions of Employment Roles in Transport and Logistics

Background

An understanding of the nature and evolution of student perceptions of employment roles in shipping and logistics will surely both enhance and precede assessment of the role of employment and other motives in the decision to undertake postgraduate study in these subject areas. This understanding is needed to:

1. Enable teachers to build on and develop existing student knowledge.
2. Review any characteristic attractions of particular occupations.
3. Define the role of such perceptions in attracting students into vocational courses.

With this in mind, this chapter reports on two studies that traced how student perceptions of employment roles in transport and logistics evolved. These studies help to define a methodology for conducting similar work and also reveal issues that were important to students when thinking about vocational employment in transport and logistics generally.

The first study aimed to explore how perceptions of the functions of transport planners and related occupations developed in full-time undergraduates enrolled on transport courses, raising questions of how perceptions of employment roles might be modelled. The views of students engaged in transport planning studies on Stages Two and Three of their courses (TR2 and TR3) are compared with those of freshmen who had not yet made this choice (TR1), and finalist Rural Resources Management students (RR3) who were studying some transport planning. The approach offers insights into the development of a schema of perceptions, which may also influence the decision to choose the subject.

The second study reports how student perceptions of several occupations in logistics and distribution were observed to evolve. It yields evidence of schema development and insights into how a teaching package could be developed to accelerate the student awareness. Features of employment roles that might attract potential recruits into these professions at a marketing level are also raised.

Transport Planning and Transport Planners

In the light of its recent rejuvenation, transport students' perceptions of transport planning as a profession should be relatively well informed. Equally, recent attempts to establish a dedicated professional body for transport planners (The Transport Planning Society, 1997) imply that deeper professional identities should exist, which define this particular group of employment roles. However, little is known about the ways in which student understanding of such identities evolve (Dinwoodie, 1994).

In this study, a qualitative approach was adopted. The context involved requesting students enrolled on transport courses to describe their perceptions of some employment roles which they might seek to undertake after having decided to study transport planning. This is similar to situations where postgraduate university applicants, continuously evaluating whether to continue in employment or return to study, might be requested to describe their perceptions of possible future roles, which they might seek to adopt after graduating from advanced courses.

The discussion begins by considering approaches towards defining the features of transport planning as a profession and how transport students have been observed to describe transport employment roles. Activities as highlighted in the Careers literature, which define the selected employment roles, are also described. The next section outlines the study methodology, and categorisations of the activities associated with eight employment roles. Roles included the transport planner (TrP), consultant transport planner (CTrP), town planner (ToP), highway engineer (HE), chartered civil engineer (CE), development control officer (DCO), transport modeller (TM) and road safety officer (RSO). Two researchers (Res A/B) analysed the results. The evolution of the set of perceived activities of these roles are described along with a schema of concepts, within the context of the decision to study transport planning.

Engineers, economists, town planners, transport operators and other groups might all make claims to owning the heritage of transport planning, but two issues likely to feature in student perceptions involve distinctions between strategic and tactical issues, and the terms 'transport' and 'transportation'. The first issue distinguishes between decisions involving policy or strategic matters on the one hand and detailed or tactical matters on the other. The former generates development plans dealing with the broad physical structure of an area and the policies and priorities for its future development, while the latter concentrates on specific allocations and details of implementation. The second issue differentiates between 'transport' as the 'means or act of conveyance' and 'transportation' referring to 'the whole system of determining why, where and when people and goods are conveyed' (The Institution of Highways and Transportation, 1997, p.iii). Students interested in the business of transport, might be expected to associate this more with the tactical activities of a transport planner employed in a transport operating company responsible for scheduling and managing cargo movements on a regular basis. They are probably less likely to think of more strategic actions involving the design or management of the infrastructure of transportation systems.

Early work into how students describe transport related employment roles (Dinwoodie, 1994) defined elements relating to the actions, function and content of a role. The action element sometimes includes phrases such as 'responsible for', 'oversee' or 'manage' or generic verbs such as 'plan', 'control', 'decide', or 'advise', and is present in most descriptions. The functional element is less prevalent, referring to more precise actions such as where a generic activity of 'planning' might include functional roles in 'routing', or 'scheduling'. The content of the role is variable and can refer to geographical areas, particular firms or a whole industry, or parts of network ranging from one link to a corridor. By way of illustration, we shall concentrate on the action element of the eight role descriptions, which are summarised in Table 3.1 as outlined in a sample of the careers literature which may be familiar to undergraduates making these descriptions (e.g. Goetz, 1990; Leavesley, 1995; Transport Planning Society, 1997). Neither the sources nor the categorisations are exhaustive, and no functions or contexts are shown, but they indicate some of the stimuli which students may have been exposed to.

Table 3.1 Activities shown in the careers literature that describe selected roles

X denotes that an activity is described in the careers literature

Activity	TrP	CTrP	ToP	HE	CE	DCO	TM	RSO
Plans	X	X	X	X				
Analyses			X	X				X
Solves problems	X	X						
Models	X	X					X	
Forecasts							X	
Assesses			X			X		
Controls						X		
Deals with				X	X			
Designs	X	X	X		X			X
Maintains								X
Decides			X					X
Engineers					X			
Manages	X	X	X		X			
Coordinates			X					
Works out			X					
Studies			X					
Educates								X
Helps		X						
Assists		X						

Source: the author

Data Collection and Perceptions of the Transport Planner

Undergraduates enrolled on transport and related courses at the University of Plymouth were asked to describe on one line, the work of eight transport planning related employment roles. The students surveyed included all those present in each of four classes. These included 41 sitting an introductory course in Transport (TR1) about one third of whom were likely to major in Transport, 18 TR2 and 14 TR3 students, whom had decided to study at least some Transport Planning, and might proceed to careers in the area. In addition, 13 RR3 students studying Transport as part of a course in Planning were included. Both the course stage and student perspectives on transport planning could influence their responses.

Responses were analysed qualitatively using content analysis, which provided a useful means of analysing loosely structured or open-ended questions (Millward, 1995). The advantages of enabling respondents to express answers using the concepts and language which they felt most comfortable with were retained, and in the absence of prior knowledge of the concepts likely to be expressed, ensured the content validity of the study. When a response was recorded, either its meaning could be analysed subjectively (qualitative content analysis), or the frequency with which some event occurs could be recorded, including that of particular words, phrases, concepts or physical events. In this study, a quantitative analysis of the occurrence of particular phrases and concepts in the student descriptions were noted by two researchers, A and B, drawn from different professional backgrounds. Both conducted independent and blind quantitative content analyses of responses for each role, and both identified dimensions relating to the perceived actions of each occupation and the contexts within which they were practised, but noted few references to functions. Two items each for the activity and context elements were sufficient to summarise most responses, with the second often proving to be redundant.

In selecting statistical procedures with which to analyse the data, it was necessary to test whether both researchers were using the same process to assign responses to particular categories of activity, before testing for differences in the frequency distributions obtained between the four subgroups of students. Table 3.2 shows the results of an initial classification of the Transport Planner's role by Researcher A, which generated 20 categories of activity and 19 categories of subject context. When extended to all eight roles, 36 categories of activity were generated, for just one researcher. In addition to high non-response rates for the Development Control Officer and Transport Modeller roles (Table 3.3), descriptions such as that of a Transport Planner who 'plans transport' (Table 3.2), indicate unsophisticated schema in many students.

Problems were encountered in comparing the classifications of the two researchers. The raw categories of activity assigned to each response, based on nominal scale data, were not suited to the usual measures for assessing the reliability of scores assigned to the same object between two different assessors (e.g. Cramer, 1997). The solution adopted involved using binomial measures to compare pair-wise assignments between researchers to a particular activity for a given role.

Table 3.2 Activities and contexts of the transport planner as categorised by Researcher A

The number of cases in each category are shown

Activity		Subject Context	
Plans	51	Transport	28
Designs	8	Routes	14
Develops	7	Transport systems	11
Analyses	4	Roads	9
Works out	4	Transport network	7
Decides	3	Geographical area	5
Models	3	Infrastructure	5
Assesses	2	Schedules	5
Builds	2	Traffic flows	4
Controls	2	Public transport	3
Co-ordinates	2	Road network	3
Ensures	2	Transport needs	2
Is responsible for	2	Transport operations	2
Organises	2	Transport policies	2
Solves problems	2	Transport options	1
Advises	1	Congestion	1
Deals with	1	Constraints of transport	1
Forecasts	1	Demand for travel	1
Improves	1	Highway	1
Lays out structure	1	One mode	1
Schedules	1	Projects	1
Restructures	1		
Total responses	103	Total responses	107

Source: the author

Table 3.3 Frequencies of activities perceived for each role

Activity	TrP	CTrP	ToP	HE	CE	DCO	TM	RSO
Develops	5	2	3	4		1	2	
Plans	49	7	32	11	8	1	4	1
Analyses	4	2	1	3		1	4	2
Solves problems	2	2	1	2			1	1
Models	3	1		2			37	
Forecasts	1						7	
Assesses	2		2			2		4
Controls	2		4			25		2
Deals with	3	1	2	2	2			6
Designs	8	1	6	22	22		6	2
Maintains			1	12	1			4
Lays out	2		13	4	2			
Decides	1		1	3			1	2
Draws			3		2			
Engineers				4	5			
Manages			2		2	3		1
Inspects				7	4	6		8
Co-ordinates	2	2	3		2	3		1
Implements	2			1		1	1	1
Organises	4	2	3			1		
Advises	1	35	2		3	3	1	4
Ensures	2		1	2	1	3		14
In charge of				1	1	2		5
Works out	3		1	1				
Studies / looks at		6	1	1	1	1	2	1
Regulates						6		2
Makes			2	2			2	
Educates								10
Monitors			1	3	4	8		5
Improves	1			3			1	9
Responsible for	2		2	9	1	3		9
Helps / assists		13	2		1			31
Builds	1	1		17	6		1	
No response	5	7	6	2	14	19	19	4
Missing cases	0	10	4	7	16	3	1	1

Source: the author

For each role, the proportion of respondents whom Researcher A felt had identified a particular category was assumed to represent an expected proportion of successes. Each response was assumed to represent an independent trial between researchers, repeated for each respondent, and a two-tailed test of a null hypothesis of no difference between the proportions allocated by each researcher was conducted. The probability of the proportion of responses allocated by Researcher B occurring within a cumulated binomial distribution was estimated (Levine et al, 1997), and comparisons were repeated for each activity and each role, within the overall sample and subgroups. Proportions allocated by each researcher, and cases where the null hypothesis of no difference between allocations to categories at the 95 per cent (99 per cent) level of confidence were rejected, are shown as * (**) in Table 3.4.

Table 3.4 shows that the eight-point activity classification was most effective for the transport planner's role and least effective in engineering roles, but classifications for a majority of role activities were consistent between researchers. There were similar results within the sample subgroups. There were problems in classifying activities in the consultant transport planner's role, where Researcher A classed consultants as being 'independent / in the private sector' or 'dealing with detail / specific schemes', and the road safety officer's role, with differences in the classification of modelling, controlling and analysis activities. In any event, the significance of differences was unstable, still being dependent, at least in part, on factors such as the number of items in the taxonomy and the number of active categories in any one distribution, and the observed frequencies within particular categories.

Table 3.4 Replies allocated to eight categories of activity

Data shows the percentage of replies allocated by Researcher A and by Researcher B respectively as 6, 6
Researcher A allocated 35% of replies to 'other' by for CTrP

Activity ALL n=86	TrP	CTrP	ToP	HE
No reply	6, 6	35, 21	16, 12	3, 10**
Develops	9, 12	0, 2	19, 26*	36, 36
Analyses	13, 17	22, 21	8, 17**	15, 19
Models	13, 15	40, 52	6, 10	8, 12
Controls	1, 5*	0, 0	43, 7**	5, 14**
Designs	58, 63	0, 10	0, 41**	8, 15*
Maintains	5, 2	0, 0	2, 0	31, 20**
Organises	0, 0	0, 0	0, 1	10, 15
Builds	8, 5	3, 1	1, 2	1, 5**
Differences	1	-	4	5
Activity	CE	DCO	TM	RSO
No reply	17, 37**	22, 26	22, 23	6, 6
Develops	44, 36*	0, 0	7, 7	0, 38
Analyses	15, 9*	26, 15**	1, 2	58, 24**
Models	3, 6	6, 9	26, 13**	9, 36**
Controls	6, 10	47, 55*	0, 0	29, 1**
Designs	8, 8	0, 2	0, 7	0, 0
Maintains	12, 6*	0, 0	0, 1	0, 10**
Organises	1, 2	0, 2	0, 0	0, 0
Builds	0, 0	0, 0	45, 53*	0, 0
Differences	4	2	2	4

Source: the author

Perceptions of Related Roles

Spearman's rho correlation coefficient was used to measure the association between the ranked frequencies of each category in the subgroup distributions analysed. Tables 3.5 and 3.6 summarise the computed correlation coefficients, with values exceeding + / - 0.60 being sufficient to indicate that the nine pairs of frequencies are likely to be drawn from common parent distributions, at the 95 per cent level of confidence. For most roles, excepting those of the town planner and road safety officer, there is 95 per cent certainty that the subjective classification processes used by the two researchers generated statistically significantly correlated sets of results. Similar findings apply to the analysis of the subgroups of transport planners and development control officers, but not to classifications of highway engineers. In the latter case, even given the overall correlation of 0.96 between the two distributions, pair-wise items within these distributions were much less well matched. Where correlations by the same researcher varied within the subgroups, this provided evidence of schema development. Even for the development control officer's activities, where students' views were the most stable, neither researcher found any significant correlation between TR1 and RR3's views. For the transport planner's role, neither the overall view nor those of any other subgroup were correlated with TR2's responses. The highway engineer's role also generated distinct schema differences where apart from some links between TR2 and TR3, there were no significant associations between subgroups. Cross-role correlations of perceived activities were also attempted for related roles, including links between transport and town planners, highway and civil engineers and development control and road safety officers. These confirmed the correlations between each pair as perceived by the RR3 subgroup, but added little concrete information. For Researcher A, the overall correlation between transport and town planners was negative, and there was no overall correlation between development control and road safety officers for Researcher B, despite good correlations within the subgroups. The lack of correlations among TR2 and TR3 groups regarding engineering roles, indicate an unstable schema.

Table 3.5 Correlations between roles, researchers and course stages

Data shows Spearman's rho correlations between classifications of Researcher A and Researcher B as 0.81, 0.84.

Correlation	TrP	CTrP	DCO	HE
Res A / Res B	0.94	0.811	0.91	0.96
TrP with:	All	TR3	TR2	TR1
TrP,all	0.94			
TrP,TR3	0.81, 0.84	0.77		
TrP,TR2	0.53, 0.56	0.46, 0.42	0.90	
TrP,TR1	0.79 , 0.89	0.43, 0.72	0.30, 0.23	0.89
TrP,RR3	0.75, 0.82	0.55, 0.79	0.04, 0.03	0.83, 0.85
HE with:	All	TR3	TR2	TR1
HE,all	0.96			
HE,TR3	0.77, 0.66	0.52		
HE,TR2	0.91, 0.65	0.76, 0.24	0.51	
HE,TR1	0.21, 0.95	0.04, 0.54	-0.03, 0.54	0.35
HE, RR3	0.42, 0.47	0.54, 0.19	0.40, 0.57	0.33, 0.38
DCO with:	All	TR3	TR2	TR1
DCO,all	0.91			
DCO,TR3	0.80, 0.90	0.80		
DCO,TR2	0.99, 0.96	0.80, 0.92	0.93	
DCO,TR1	0.94, 0.99	0.58, 0.88	0.94, 0.95	0.93
DCO,RR3	0.80, 0.64	0.99, 0.71	0.80, 0.52	0.58, 0.59

Source: the author

Table 3.6 Correlations between roles, researchers and courses

Data shows Spearman's rho correlations between classifications of Researcher A and Researcher B, as 0.81, 0.84.

Correlation	TM	ToP	CE	RSO
Res A / Res B	0.87	0.23	0.84	0.35
TRP with:	RR3	ToP		
TrP,all		-0.11, 0.89		
TrP,TR3		0.55, 0.62		
TrP,TR2		0.39, 0.24		
TrP,TR1		0.87, 0.93		
TrP,RR3	0.93	0.91, 0.89		
HE with:	RR3		CE	
HE,all			0.50, 0.22	
HE,TR3			0.41, -0.01	
HE,TR2			0.05, 0.26	
HE,TR1			0.09, 0.33	
HE, RR3	0.68		0.58, 0.54	
DCO with:	RR3			RSO
DCO,all				0.96, 0.00
DCO,TR3				0.77, 0.56
DCO,TR2				0.99, 0.90
DCO,TR1				0.86, 0.82
DCO,RR3	0.65			0.75, 0.69

Source: the author

Towards a Schema of Planning Related Roles

Transport students' perceptions of the transport planner's main activity varied from 'design' (TR1), through 'modelling' (TR2), perhaps reflecting course nomenclature, to 'developing' (TR3) and 'analysis' (RR3). Compared with perceptions of the town planner's role, the 'design' element was weaker early in the courses, but persisted in more advanced students, although Transport finalists perceived more 'analysis' and Rural Resources students more 'development'. In describing consultants, Researcher A used a distinct taxonomy, but the concept was rarely manifested in students who had not yet chosen to study transport planning. Similarly, freshmen were unfamiliar with the modeller's role, and although more students on transport planning courses used the verb 'model', or identified the concept of 'building', this alluded to both mathematical models and construction activity. Chartered civil engineers were perceived as being engaged in 'developmental' activities, with some disagreement in classification between the design and control elements, but a wider range of activities including maintenance issues, were attributed to the highway engineer. Road safety officers were associated with more analysis and control activities, but development control officers, seen almost exclusively as representing control issues by the rural resources students, were also perceived as having an analysis function by transport students. The role was not well understood, overall. Accurate perceptions of the roles of consultants, road safety and development control officers, modellers and engineers only developed after the decision to study courses in transport planning had been taken.

In this analysis, statistical measures comparing the perceptions of subgroups of students confirmed some of the problems of classifying attributes between researchers and the existence of statistical variation between groups of students, but did little to explain how a schema of concepts might develop. Few robust statistical techniques were readily available with which to make inferences about pair-wise nominal-scale comparisons between small numbers of individuals, rather than frequencies, in the group distributions. In the end, the adoption of an eight-point taxonomy in the content analysis of role activities was a subjective decision with few statistical techniques available to justify it. The analyses of role contexts, not reported here, were even less amenable to such treatment. If employment considerations may be significant to prospective shipping and logistics students, then the ways in which similar students view employment roles, and how their perceptions evolve, is fundamental to understanding this relationship. In this study, quantitative content analysis, based on subjective processes, provided a useful prelude to broader statistical analysis, best reserved for detailed analysis of a well defined tick-box style instrument.

Changing Perceptions of the Role of the Logistics Analyst

The second study aimed to investigate whether the rate at which students develop an understanding of career openings in intermodal distribution could be raised. This work was stimulated by concerns that the concept of 'logistics' and the role of an 'analyst' were not well established even in second year (TR2) university students. There have been useful attempts to define particular roles in logistics (e.g. Murphy and Daley, 1997), but the ways in which student knowledge of these roles is acquired is not widely reported. As a first step, it was necessary to establish the existing levels of student knowledge at various stages in courses, before the existing teaching regimes could be reviewed to make learning more effective. Some of the 'softer' aspects of logistics, such as those relating to ethical education, are known to be difficult to teach or may involve longitudinal teaching spread over several semesters (Daley, 1994). This approach is consistent with the ideas of educators who identify a learning process that may involve learning through a series of steps. These begin with sensory experiences that generate information, which once processed by the brain, commence the process of learning, which is later reinforced. At their lowest level schema may simply be facilitating the recognition of regular patterns of experiences. At a higher level these experiences, taken as a whole may be organised before being retained as knowledge. Because an experience is complete, this implies that if we see a pattern, it is associated with a concept. In this way, if a nametag is attached to a face or a melody is experienced, this tags a concept, in turn tagging a pattern represented by an abstract, image-less and wordless element of thought.

By asking how a student develops an understanding of employment roles in intermodal distribution, we are attempting to reconstruct and explore their concept structure. This assumes that words can tag the concepts in the structure, so that a cognitive map is defined, where the territory of verbal concepts, have been recorded on paper. Nonverbal concepts only become expressible when a socializing experience results in labels understood by at least one other person being attached to them and meaning requires at least two sensations to be mapped.

As their subject knowledge grows, university logistics students should eventually raise their familiarity with relevant occupations, to the point where trained researchers could unearth developmental stages within their understanding. Once the extent and nature of student knowledge, at various stages of the existing study program, in for example Transport at Plymouth could be traced, this would enable areas that would benefit from changes in understanding, to be identified and prioritized. A questionnaire was devised to reveal the concepts that students were using to describe their understanding of a range of occupations in logistics, and hence to reveal the ways in which their concept structures in this area were changing. The questionnaires were then administered to whole classes. This approach ensured that simultaneous replies were received from students in each group and overcame some of the problems of conducting interviews, which demand longer time frames. As such, if responses were tainted by more recent experiences such as lectures or field visits, they would probably have affected whole groups, with less random impacts on individual replies by such unstable influences. The nature of any schemas that emerged from this research, were designed to be of interest to logistics teachers, rather than diagnostic tools for use by vocational guidance professionals. The latter are less interested in the levels of knowledge of students about particular employment roles, but more in skills competence for career management in 'exploring resources, reflecting on past and present, planning, monitoring and evaluating self and situation and developing autonomy' (Kidd and Killeen, 1992, p.230).

In order to focus their attention, students had already been asked to answer questions relating to their own preferred employment within the transport industry, reasons for their choices, sources of information they had used, and details of their previous industrial experience. Their reasons for choosing to study Transport at university, and in particular at Plymouth, were also explored (Dinwoodie, 1996). At this point, students were asked to provide one-line descriptions of the work involved in various occupations in international distribution selected to include several modes and distribution functions with a bias towards logistics. Logistics was a subject of interest to some students sampled in a control group, all of whom had chosen not to major in Transport, and indeed not to study any specialist Transport at all.

In designing sampling procedures, a comparative study of university students at the same stage in their careers was attempted, between those studying Transport (TR1 and TR2) and non-transport stage one students, acting as a control group. All Transport students present in relevant classes were surveyed in the first week of term to prevent any bias from current teaching. One hundred per cent sampling rates, using questionnaires administered to groups without warning and collected immediately with no exchange of ideas between students, provided unpremeditated first impressions from the following groups:

1. A control group including Geographers and Maritime Business students, who had chosen not to major in Transport, but who might reasonably have done so.
2. Thirty eight TR2 Transport specialists including some international exchange students new to Plymouth but with prior Transport education

and work experience, and 'single honors' or 'major' students from the Plymouth TR1 program.

3. Thirty TR1 students, including some whom may opt for major or full degrees in Transport, and others whom may select related Maritime Business or other programs.

The research used open-ended questions, designed to reveal concepts considered significant to those answering, and content analysis of replies (Breakwell, Hammond and Fife-Schaw, 1995) which allowed sufficient commonalities for some statistical comparison, although this was not a high priority in the empirical approach adopted. Null hypotheses of no significant difference between the proportions stating a given attribute in different groups were tested against a one-tailed alternative hypothesis of a greater (or lesser) proportion p, using Z tests of pooled proportions. Where small samples n denied its use (where $n.p < 5$), tests were not attempted, as the inferential power of binomial enumeration is low.

Non Response

Where students failed to reply, this represents a lack of awareness in terms of schema development (Boreham and Arthur, 1993). Response rates to questions are shown in Table 3.7, with TR2 recording lower non-response rates for all jobs, indicating increasing awareness after one year of study. The control group displayed high non-response rates in relation to the roles of distribution manager and logistics analyst, but fared better for freight forwarder, and similarly to TR1 for other roles. The distribution manager's job was the best known role among Transport students, with rail and logistics analyst jobs least well known. Analyst roles and freight forwarder were not described by a majority of TR1, but 20 per cent more of TR2 were aware of these, with 15 per cent more for shipbroker and one per cent for distribution manager. In terms of schema development, awareness of managerial functions developed ahead of planning, with technical concepts such as 'logistics', 'marketing analyst' and 'broker' developing later.

Table 3.7 Percentage non-response by subgroup and employment role

Role	TR1	TR2	Control
Distribution manager	30	29	64
Logistics analyst	57	37	68
Freight forwarder	50	29	40
Rail marketing analyst	60	39	52
Shipbroker	47	32	48

Source: the author

A Logistics Careers Teaching Package

In order to increase their awareness of key functions in intermodal logistics, TR1 were presented with a package of ad hoc learning activities. In an introductory lecture, they were asked to discuss official statistics showing recent trends in employment in the industry by mode of transport, and detailed occupational and industrial categories. Next, support staff from the Careers Advisory Service introduced students to the concept of self-awareness, and possible types of relevant employment and sources of information available for exploring employment opportunities. In the main exercise students were requested to form groups of three, to research sources of information for several employment roles. For each role, they were asked to write job descriptions of about 100 words and show the addresses of five relevant organizations, and then list and briefly describe five other jobs that a person in each role might come into contact with during the working day. They were then requested to list the educational requirements needed to perform each of the roles shown, and to record fully all information sources used during the exercise. They were then required to present a one-page report on each role that could be duplicated and shared with the rest of the class, either orally or in writing. Specialist careers staff assisted in finding relevant sources of information and evaluating these sources. Finally a 'value-added' survey was conducted, which involved repeating parts of the original questionnaire, to highlight any changes in responses.

Non-response following teaching fell to 3 per cent for distribution manager, freight forwarder and rail marketing analyst roles, 6 per cent for shipbroker and 9 per cent for the logistics analyst, reflecting a substantial increase in knowledge following relevant teaching. Statistical comparisons between group proportions of non-response indicated 95 per cent certainty that proportions following teaching were drawn from different populations compared with those before teaching.

Elements of Employment Roles Identified by Students

A content analysis of the one-line job descriptions indicated action, function and content elements associated with some roles, but fewer elements in other roles. The job function and content elements identified were specific to each role, but the action elements of some roles were more general. Detailed findings for each role are presented below. The 'action' element of job descriptions revealed categories of:

1. Responsible / make sure that
2. Manage / oversee / coordinate
3. Contact with customers
4. Control
5. Plan
6. Decides
7. Study / find / investigate
8. Optimize / advise.

Responsibility implies a board level function, but management a lower level, and control or customer contact could be at either level. Planning, deciding, studying or advising imply a horizontal staff function within the organization.

The Distribution Manager was misconceived in terms of his actions, with some initial confusion between the planning and executive management actions amongst stage one students, which had been corrected by TR2. Some ten per cent of TR1 wrongly perceived the distribution manager as 'planning' or 'deciding', compared with all of TR2 who responded, who identified him correctly as solely 'managing' or 'being responsible'. After the teaching package, the proportion of TR1 stating 'manage / oversee / coordinate' actions rose from 23 to 74 per cent, a statistically significant shift. This contrasted with only 24 per cent of the control group who were aware of the functions of the distribution manager; most failed to reply.

The function of the distribution manager was described initially as 'organizing operations, routing / scheduling' or 'handling', by 37 per cent of TR1, but by only a couple of TR2. More sophisticated concepts relating to 'distributing', 'how to transport', 'inflows and outflows' and 'movement and storage' were reported by TR2, and a few of the control group who did reply. The content of the distribution manager's job was identified as 'areas / places' (13 per cent of both TR1 and TR2), but the 'firm' or 'distribution firm' (TR1) and 'products / goods or processes' (TR2) were also noted. This reveals a greater generality of understanding in TR2, but was also the case for those in the control group who did reply. After the teaching package, TR1 perceived the function as 'organizing operations' or 'distributing / delivering', and the content as a 'distribution firm' or 'products / goods'.

Typically TR1 students defined the distribution manager as somebody who 'manages (distribution) operations for a firm', TR2 thought that he 'manages distribution of products', but the control group failed to reply. Even in this one example teaching was highly effective, by emphasizing differences in roles within a group, hierarchies within organizations and groups in organizations.

Logistics Analysts presented the least known role, with only 43 per cent of TR1 responding initially, but doubling following the teaching package. The median initial response of an 'investigative' role (23 per cent of TR1 and 26 per cent of TR2) rose to a statistically significantly different proportion (65 per cent) following teaching in TR1. The 14 per cent of TR1 who incorrectly attributed a 'managerial' or 'responsibility' role fell to 8 per cent in TR2, but the correct action, was identified by 50 per cent of TR2 and only 29 per cent of TR1. The concept of planning clearly developed for many students during TR1.

In terms of the function of the logistics analyst, 58 per cent of TR2 correctly identified 'routing / scheduling' or 'how to transport goods', compared with 36 per cent of TR1 before the teaching package and 65 per cent afterwards. Gratifyingly, none of TR2, but 7 per cent of TR1, initially quoted the executive function of 'distribute / deliver', and 8 per cent of the control group 'organizing operations'.

Initially 34 per cent of TR2 identified 'products / services' as the content of the role, but only 16 per cent of the control group and 10 per cent of TR1 identified the same categories, with 80 per cent initially not referring to any content context. After the teaching package, there was an increase in awareness, with role contents including 'the whole firm' (23 per cent), or 'service, processes or systems' (26 per cent), or a 'distribution firm', 'industry' or 'products' (10 per cent each).

Typical definitions for both groups of stage one students involved no reply, changing to 'studies routing / scheduling of products' for TR2 or 'studies routing / scheduling in the whole firm' following teaching in TR1.

Freight Forwarders roles were described in terms of combined elements including action, function and content. Forty per cent of TR2 students identified the concept of a 'middleman for cargo exchange' but none of TR1 did so. Instead, they referred to such concepts as 'generally seeking freight for a company'. 'Handling and planning of freight' were identified by 25 per cent in both groups, but 64 per cent of the control group failed to reply, and a majority of those who did so, confused the role with that of the distribution manager. The 'middleman' concept discriminated clearly between TR1 and TR2 students initially, but after the teaching package, 62 per cent of TR1 identified this concept, although the idea of 'planning the movement of freight' still required refinement, as the next most frequent response.

Rail Marketing Analyst roles were felt to include both a familiar marketing function, and a less familiar analyst's action. Not surprisingly, a majority of TR1 students initially failed to respond to this role. The concepts of 'market research' or 'product promotion' predominated for those who did respond. A few TR2 students (13 per cent) referred to 'statistical analysis' or 'modal competition', but some TR1 students (14 per cent) commented on 'timetable' functions. The freshmen control group concentrated responses on 'promotion' and 'statistical analysis'. Typically, responses before the teaching package involved no reply for both TR1 students and the control groups, but 'researches / promotes / advertises knowledge of customer wants' for TR2 students. Following the teaching package, a statistically significant proportion of TR1 responses shifted to 'researches trends in the market' with 'knowledge of customer wants' as the second major element, pointing to the assimilation of concepts in marketing studies.

Shipbrokers were described in terms of their action and functional elements, but not their contexts. Functions identified included those of a 'middleman' and 'finding the best deal for a customer', but a majority of both TR1 and TR2 students identified no function. Many TR2 students (47 per cent) correctly identified 'chartering ships or ship space' as the prime action, but fewer freshmen (30 per cent) did so, with some referring to the 'control of shipping' and 'the buying and selling of ships'. Amongst the control group in stage one, partial replies included the 'middleman' and 'buying and selling ships' most frequently. Typical initial responses for both Transport groups were 'a middleman who charters ship space'. Following the application of the teaching package, statistically significant changes in the proportions of students noting the function of the shipbroker as a 'middleman, dealer, agent' and 'one who charters ships or ship space' were recorded, replacing those who previously did not know.

Conclusion

The ways in which students began to acquire a knowledge and understanding of key employment roles and functions within intermodal distribution and transportation have been reported. On the basis of such findings, teachers are better placed to identify and prioritise shortcomings in their own students' understanding, and devise new approaches to teaching which can assist learning processes.

One schema was found to evolve from an initial understanding of simple, executive line management actions, as performed by the distribution manager into an understanding of planning or middlemen roles. In another area, student use of technical concepts such as logistics or marketing, and relational concepts such as competition, were indicators of a more advanced stage of awareness. In terms of the perceived content of jobs, there was evidence of a shift from 'the firm' initially, through 'product' and 'place', and eventually to 'service', and the knowledge of techniques including routing, scheduling and statistics increased in more advanced students. In the light of these findings, early teaching in Transport needs to concentrate on planning, analyst and freight forwarding or agency functions in logistics, rather than the traditional line functions. The scope for earlier exposure to basic marketing concepts, case studies and teaching of particular techniques may also be important, but no single measure in isolation appear to be sufficient to stimulate student recruitment into Transport and Logistics programs.

High levels of non-response in the stage one control group imply that only those students who are already planning careers in logistics, or those with friends or relatives involved in such work, had any real awareness of these occupations. In order to raise the knowledge and awareness of a wider range of young people in relation to these occupations, they need to be placed in situations where they must confront their future occupational selves. Logistics careers information is becoming more accessible (e.g. Rogoff, 1999), but many young people have still not been exposed to such sources, early in their careers. Practitioner assistance in providing specialist lectures, library materials, or realistic group-work exercises in which students could explore their self-awareness is essential, along with hands-on industrial work experience for young people.

These findings also have implications for the subsequent analysis of individual perceptions of the decision to embark on advanced courses of study in shipping and logistics. Firstly, the underlying theory of an evolving hierarchy of perceptions of relevant concepts, which in this chapter were related to employment roles, is both important and useful. This in turn implies a need to employ specialist cognitive techniques, such as cognitive mapping, in order to structure and understand individual study decisions. Finally, because of the complexity of tracing the ways in which the understanding of even one individual develops, it will be necessary to find ways to both standardize and then compare such maps in order to analyze differences between them. Quantitative comparisons between maps will be required in order to render this problem tractable.

4 Qualitative Analysis of the Study Decision

Introduction

The purpose of this chapter is to report on a qualitative analysis of the processes and the factors which were important to applicants whilst making their decisions to study. This exploratory empirical work was conducted within groups of mainly international students seeking to embark on postgraduate taught vocational courses in international shipping and logistics at the University of Plymouth. The accounts of their individual decisions to undertake advanced study, should be viewed within a wider educational context. On the one hand, there are increasing pressures on universities to increase both their intake and quality of provision for such students, and on the other, a paucity of research into the perspectives of this particular group, which were discussed in Chapter 2. In order to highlight the most significant issues of concern to students, qualitative analyses based on the use of focus groups and loosely structured group interviews were undertaken. The choice of these particular techniques empowered students to express the issues that they had considered to be significant when making their decisions to study, in their own words.

Methodology

This chapter, which describes work centred on vocational courses, explores a dimension of higher education relevant to the workplace, by considering the implications of student opinions for educational provision in this area (Brennan, Kogan and Teichler, 1996). By concentrating work at a single institution, detailed analysis of the opinions of individual international students was possible. Akin to Harris (1995) approach, it relates findings from one institution, which inevitably deny their wider currency, along with any claim to purport to represent systematic study of any one facet of the issue. However, in so far as that the particular combination of courses available, and the geographical attributes of any institution are unique, such research should be capable of capturing the idiographic dimensions of individual decisions which may be significant but less apparent in more broadly based surveys. Even so, some fundamental methodological issues remain to be addressed.

The complexity of this issue may help to explain the paucity of previous attempts to tackles it. As Brennan et al (1996, p.20) noted:

> International comparison ...has remained confined to quantitative trends in student enrolment and graduate employment. ...Besides the costs involved, the language barriers, the difficulties of acquiring sufficient experience, and different concepts of the tasks and functions of higher education and the character of graduate work, discourage comparative research in this area.

Or again, as Harris (1995, p.81) noted:

> Given ...the increasingly crucial economic role played by overseas students in the university system, there is a surprising dearth of recent research or writing in relation, either to their experiences and attitudes, or to their learning needs and how these may best be met...

Harris proceeded to propose, that in the absence of systematic study of the learning experiences of different categories of students, varying by such factors as nationality, culture, native language, mode of funding, gender and family status, a case study of 11 alumni from his own university might stimulate further work. Underlying this approach was an attempt to initiate an iterative debate into the constant re-matching of particular groups of student needs, and the processes of allocating resources in universities. This approach is a partial response, related to the student perspective of Harris' 'prior to arrival' stage of the life cycle approach to the overseas student. However, it only addresses the later stages of the cycle and any implications for such issues as to how to approach the problems of practical course administration or devising learning strategies for such groups, are merely raised en passant.

Any researcher attempting to analyse the ways in which multinational groups of students in international shipping and logistics use English words to communicate their thoughts, needs to be aware of several influences on the process. Firstly, the distinct functional business culture in logistics, involving the management of resources in the supply chain, including procurement, manufacturing, distribution, waste management, maintenance and project work, is broader than the maritime backdrop of shipping. Both disciplines have evolved their own jargon, but sufficient commonality of meaning has been preserved to enable cross-functional dialogue. Secondly, cross-cultural influences on groups of multinational students, and the complexities of communicating in English acquired as a second or third language, may distort the meaning of concepts to particular individuals, tainted by other concepts in a range of mother tongues. Any research methods employed to analyse how such students make decisions must enable the richness of the phrases, concepts and constructs of individual respondents to be captured. This survey aimed to elicit the phraseology and concepts used by particular groups of students, to establish the range of issues of interest to them, and their frequency, in order to establish their main areas of concern and eliminate purely idiosyncratic issues from the discussion. These considerations, coupled with minimal prior work in related areas (Dinwoodie, 1996) limited data collection to open-ended oral interviews and focus groups.

Very recently (Lane and Kahveci, 1999), studies within the field of maritime sociology have proposed a broader range of qualitative research techniques. In an investigation of transnational shipboard communities, suggested ethnographical sources for investigating cultural influences have included life histories and reflective diaries alongside focus groups and interviews. In addition, observational studies that involve participants and non-participants as well as quantitative studies of crew composition are proposed. Whilst such approaches offer exciting scope for future research into the social aspects of maritime education, they are perhaps of less direct moment within the immediate context of investigating more cognitive elements of individual decisions to study.

Bearing this in mind, interviews were conducted in late 1996, amongst two groups of post-experience or postgraduate Diploma students, and a combined group of students in the first stage of Masters programmes in International Shipping and International Logistics at the University of Plymouth. In total, seven audio taped interviews varying in length from 50 to 80 minutes were undertaken. One Diploma group included 14 students from a post-experience Diploma in International Shipping and Logistics, a one-year course for students mainly from non-European countries, with ten of them represented in two focus groups. The views of a second group comprised of 20 European postgraduate students undertaking one period of study in England, with tuition given in English, before returning to complete their studies at a French institution, were represented by one focus group of seven students and group interviews with two others. Finally three groups of students from the combined Masters programme were analysed, including mixed samples (of six, four and two students) of both European and non-European students drawn from three logistics and 11 shipping students. Overall, the views reported included 31 of the 48 registered students, and excluded only those individuals who failed to attend particular sessions. A sample summary of discussions with one of the Diploma groups, and a transcript of discussions with one of the Masters groups, are given in Appendices 1.1 and 1.2.

Content analysis was used to analyse this data (Millward, 1995), drawn from a mix of focus group interviews and group interviews (Crabtree, 1993; Morgan, 1993). Additional prompts were used in conducting these focus groups, and a copy of the prompt sheet used by the interviewer to guide each focus group is shown in Figure 4.1.

Focus Group prompt sheet

Opening statements -
I am researching the decision to study shipping and logistics at postgraduate level. All details are confidential but I may need to reproduce a transcript of some or all of your replies anonymously in an appendix to a publication: how would you feel about me doing this?

The *main question* is:
Please think of yourself in relation to postgraduate level study in shipping or logistics at Plymouth. Please discuss the issues that were important to you in this context.

Prompts may include:

1. Employment on completing the course.
2. The role of family and friends.
3. Why undertake study at postgraduate level?
4. Why study shipping or logistics?
5. Information about courses...
6. Why choose to study in the UK?
7. Why choose to study at Plymouth?
8. Where else did you consider studying?
9. Where will you live while studying?
10. Are teaching methods relevant to the decision?
11. How will you fund your course?
12. Have you any relevant work experience, and in what ways did it influence your decision to undertake study?

Figure 4.1 The focus group prompt sheet
Source: the author

The universe of material available for analysis was limited to the statements uttered, recorded and transcribed although field-notes were also used to record any other significant interventions. All stimuli were relatively response-free, but while they were unstructured in early interactions with each course group, they later became more focused to entice detailed discussion of particular issues in the decision to study. These included issues such as the perceived importance of acquiring particular personality attributes, teaching methods and funding.

Typically, sessions proceeded by asking each participant to introduce themselves to the rest of the group, before allowing free responses to the stimulus shown above. In some sessions, students were asked to consider features of their preferred employment that might attract or put them off wanting to work in it, and how their course might help them to develop particular personality traits or skills that it might require. The influence of family or friends working in the industry, reasons for studying shipping and logistics at postgraduate level and factors which might put them off, along with how they might find out about courses, were raised. Reasons for studying in the UK and specifically at Plymouth, and factors which might put them off studying both at Plymouth and elsewhere, along with any barriers to further study were explored. The results of these discussions are now reported in this sequence, verbatim where possible, and may include several points made by the same student in a particular context.

The Role of Employment Considerations in the Decision to Study

Travel and dynamism are well-known attractions of employment in transport (Dinwoodie, 1996) but under half the students interviewed in this survey (Table 4.1) raised any employment issues in the context of a decision to undertake a vocational course, and none explicitly mentioned pay as an attraction. If this is interpreted as implying that employment issues represent a course outcome, of concern near graduation, which displace earlier interest in course content, it could be consistent with Ainsworth and Morley's (1995) ex post analysis. The only features of this employment that might put students off wanting to work in it, each stated by one student, were the conflicts between continuous travel and family life, and overwork. Where 94 per cent of students failed to consider the downside, this implies a need to encourage potential recruits to be realistic in their aspirations, by offering them ongoing industrial awareness preparation classes on vocational courses.

Table 4.1 Features of employment that attracted students to it

Features of students' preferred employment in shipping and logistics that attracted them to it	% of responses
You are always learning new things. It is creative work in a dynamic environment.	26
I have a liking for ships and ports, or previous work experience.	13
It is global in nature; travel.	10
My family is in shipping.	10
You are always talking to other people and interacting with different cultures.	6

Source: the author

Only 13 per cent of students discussed the personality traits that they felt their preferred employment demanded. Two students expressed a desire to learn 'to listen to, to be aware of, and to work with other people', and 'to be more open-minded'. Also mentioned were 'personal development', 'to stand up for what you are saying', to be more 'individualistic', and an 'ability to manage one's own time'. Issues raised regarding the skills demanded of individuals, who were seeking their preferred employment, included needing an 'understanding and knowledge of the work situation'. One individual mentioned an 'ability to speak different languages' and three highlighted a need for 'knowledge of commerce or business'. Equally few students commented on the ways in which their course could help them to attain them, but a 'need for practical experience with trips to ports and practical work' concerned three students. Single responses included needs 'to live abroad and see other things', and 'for a degree in order to be able to apply to English companies'. Alternatively it would 'be good for the curriculum vitae', 'give me confidence to find a job because I don't know enough yet', 'give me specific knowledge', or create 'apprehension that I may not fit into my country when I return'.

A few students mentioned family or friends who had worked in shipping or logistics and influenced their own study decisions. This had involved 'talking things through with them' (10 per cent), or the decision had 'affected the family business' (6 per cent), their family had introduced them into shipping (6 per cent), or their parents were funding their studies.

With the possible exception of the latter, there is little evidence here to deny the view that most individuals making this particular decision to study claimed to be personally responsible for their choices, although family and friends may have influenced the precise timing of studies. The low stated priorities given to 'skills development' and high stated personal ownership of the study decision are consistent with earlier findings reported amongst undergraduates (Burn, Cerych and Smith, 1990).

Why Study Shipping and Logistics?

Students had much clearer views regarding the reasons which made the study of logistics and shipping at postgraduate level attractive or otherwise (Table 4.2). Where subject interest and novelty predominated over basic or family issues, this implies a need to emphasise both vocational and industrially specific features when designing and marketing courses. This contrasts with the emphasis placed on 'employability' in the 'cloudburst' of Masters courses (Kogan, 1994, p.63). The main reasons that might deter potential students were financial or personal, and imply a need for thorough preparation before commencing study, and substantial induction programmes at the host institution. Academic issues were raised where one student noted that 'after ten years experience, it is really hard to go back to study', and another noted that the decision could not be rushed, where 'it has to be the right course as you cannot repeat it'.

The main ways in which students had found out about postgraduate courses, included written sources of magazines, books, and brochures (23 per cent) and talking to other students (16 per cent), lecturers (13 per cent) and people in industry (19 per cent) or the British Council (6 per cent). These methods were consistent with Woodhall's (1989) approach and imply a need to ensure that university recruiters are aware of the importance of both ongoing personal contact and the potential role of high quality published material. In so far as that this study found no evidence of any consistent sequential process of, for example, an initial oral recommendation being followed by literature searches or vice versa, there are few grounds for recommending that either medium be targeted towards attracting the initial interest of potential applicants.

Table 4.2 Attractions and turn-offs of postgraduate study

Reasons which made studying logistics and shipping at postgraduate level attractive	% of responses
A need for more study, following work experience.	23
A desire to specialise in shipping.	19
A desire to change career or to keep learning.	16
A desire to learn new things - it is more interesting.	13
A desire to improve my employment prospects.	10
My family, or personal reasons.	6
Reasons which might have put students off studying logistics or shipping at postgraduate level	
The offer of a good job affording good experience.	13
Study is very expensive.	13
I need work experience in order to understand better.	6
Health or family pressures.	6

Source: the author

Why Study in the UK and at a Particular University?

Major factors which had made studying logistics and shipping in the UK most attractive related to academic issues, where comments included: 'British Diplomas and degrees are recognised around the world' (16 per cent of students) or 'it affords better academic opportunities' (10 per cent). Other factors included interest in life abroad, where 'to stay in the UK, makes life interesting' (10 per cent) or 'to practise my language' (16 per cent). Others wanted to 'study in a country well known for shipping' (10 per cent), but two students had 'family or other contacts in the UK' and one noted a 'political crisis at home'. In general, the appeal of living overseas for postgraduate students seems weaker than in undergraduate study decisions in business studies (Burn, Cerych and Smith, 1990), with academic and status issues prevailing. University managers who provide high quality and well-recognised courses appeared to be more likely to attract these vocationally committed postgraduate students.

Overall, considerations of the specific reasons which had made studying logistics or shipping at the particular university attractive, received the highest ratings. Amongst the European Diploma students, in England via an exchange programme, their reasons for choosing the French Institute offering the exchange, were paramount. These included its 'good reputation' (30 per cent of students), it was 'uniquely specialised in shipping' (30 per cent), it had a 'practical training period' (20 per cent), it offered 'a professional Diploma' and 'it looked interesting'. Similar issues featured in the decision to choose to study at Plymouth (Table 4.3), including employment prospects and status of the qualification offered, course related issues and the influence of friends or other local ties. Decisions to study at particular institutions internationally, both in this survey and for undergraduates (Burn, Cerych and Smith, 1990) were dominated by the perceived relevance of course content and quality offered. However, many aspects of these issues must surely remain in the realms of 'shared subjectivity', not easily amenable to objective measurement (Wright, 1996, p.80).

Table 4.3 Reasons why students chose to study at Plymouth

The reasons which made study of logistics and shipping at Plymouth attractive	% of responses
Plymouth degrees are recognised making it easier to find a job after the course ends.	75% of MSc
My former lecturers recommended the course.	32% of MSc
The course specialised in shipping.	42% of MSc
The course is unique.	13
I had local family ties or was attracted by the beautiful location.	13
My country does not offer these qualifications.	10
I had already studied at the particular university.	10
I had friends who had completed the course.	10
It is an entry to the Masters course.	6
I was impressed with the admissions policy.	6

Source: the author

What are the Barriers to Study?

Relatively few students raised any issues that might have put them off studying logistics or shipping at the particular university. However 'a poor quality rating or poor value for money' would deter 16 per cent, two Diploma students would be deterred if 'not offered a place on the Masters course', and one would be concerned if 'given insufficient information'. Only one Masters student considering study elsewhere gave a reason for not doing so, namely that 'the course emphasis was inappropriate', echoed by two other Diploma students. Course cost concerned two students, course length a third, and inability to get a visa a fourth. A clear institutional image, perhaps defined by organisational devices such as University Subject Panels (Pritchard, 1994, p.263) might present one vehicle for focusing overseas marketing and course identities.

The barriers that might have put students off studying logistics or shipping at Plymouth, were rarely mentioned by most of them. Two Diploma students 'couldn't afford the MSc', but one Masters student was 'willing to pay a bit more for what I want'. Other comments relating to cost included one that 'the total cost is less, when I consider the total time involved' and another that 'my company and parents partly funded and encouraged me'. In terms of preferred teaching methods, three wanted a 'practical course, not just lecture based', in similar vein to two wanting 'links with professionals'. One 'didn't know what to expect in advance' and another was impressed that 'lecturers made students welcome'. An important but lone comment noted that 'if the programme suits me, I will compromise the rest such as the city and my accommodation'. A corruption of Harris' (1995) life cycle analogy may be useful here, where failure to attract results in failure to conceive and mistreatment during pregnancy results in abortion, but surely, few foetuses consider aborting because they are apprehensive about post natal conditions.

Summary

In terms of research methods, qualitative analysis of focus group sessions and unstructured group interviews were employed. This approach ensured that the multinational respondents were empowered to define the issues that were important to them personally in making their decisions to undertake postgraduate study. The structured approaches to particular issues in more comparative contexts, discussed in the succeeding chapters, are based on the exploratory qualitative work discussed here. As an example, questions relating to particular features of students' preferred employment in shipping and logistics, initially thought to influence their decisions, were rarely raised in these focus groups. These issues may relate more to the outcomes of study than to the reasons for undertaking it, with realistic student perceptions of their future employment prospects only becoming apparent near graduation (Dinwoodie, 1996). Where few students were found to be willing to discuss an issue, such as the ways in which courses might develop particular dimensions of their own personalities, this signals a need to define the degree of importance attached to particular issues. One approach would be to rate them as relevant, important or critical.

The students who took part in these initial focus groups were attracted by both subject interest and basic employment needs, and both written and oral sources of information had influenced their decisions. They held British qualifications in high regard, but less weight was given to a desire to study abroad than would be expected for undergraduates, with specific features of a course and its reputation determining their choice of university. Course costs and a poor image, were the major barriers which could have deterred them from their studies. However, university managers who ensured that courses had a clear identity and a good professional reputation, were considered to be more likely to attract students on to international courses at this institution.

In order to test the extent of these views, it was necessary to devise and pilot a tick-box instrument based on the discussions reported in these focus groups. This instrument could then be administered to other groups and cohorts of students.

5 Quantitative Analysis of Evolving Perceptions of the Study Decision

Background

This chapter presents a quantitative analysis of changing perceptions of the decision to study at postgraduate level in international shipping and logistics (ISL) amongst groups of students at the University of Plymouth. It begins by describing a bespoke instrument developed initially from focus groups (Chapter 4), and reports on surveys conducted of the perceptions of final year undergraduates (Ug), postgraduate Diploma level students (Dp), and students enrolled on Masters courses (Ms). After briefly outlining the methodology of the study, the sample characteristics and the diachronal stability of perceptions of students in one Masters group are discussed. Some substantive results are presented in relation to the role of the choice of subject, the country, and the university of study, before considering a range of perceived barriers to undertaking study, and summarising the main findings.

Methodology

Following the piloting and testing of trial questionnaires based on analysis of the earlier focus group interviews, a tick-box instrument was devised which enabled postgraduate students to describe aspects of their decision to study international shipping and logistics at Plymouth. The decision was characterised by nine major issues, each of which were subdivided into a range of more detailed items (Figure 6.1 and Appendix 1.1). In this part of the analysis, the degree of importance which students placed on both the major issues themselves, and the detailed items within them, are discussed in relation to their overall decision to embark on postgraduate courses. Each issue or item which a student considered to be critical to the decision was scored as 3, if important it was scored as 2, if relevant as 1, and if not relevant as 0. In this section, descriptive summaries of the mean scores of particular items are presented for convenience. However, most of the statistical comparisons between groups that are presented, are based on non-parametric tests of inference. In this way, they reflect the original ordinal evaluations by students.

Most of the surveys reported in this section were conducted between October 1996 and January 1998, presenting a relatively stable period covering only two cohorts of students. The January 1999 groups were excluded because they represented a different, larger Masters group, and one which may have been responding to a changed environment, due to the additional year elapsed. All surveying was unpremeditated, and responses were collected from all students present in class when surveyed.

The groups for which surveys are reported here included:

1. Undergraduates surveyed six months prior to graduation. These included 47 students of maritime business, and 13 of transport, many of whom had reasonable expectations of postgraduate study, and some who held offers of places.
2. Diploma students including 'European' groups, studying jointly at institutions in England and France, and 'International' groups, of mainly non-European origin. Fifty-two responses were analysed, drawn from two cohorts of each group, all of whom were studying combined shipping and logistics.
3. Successive cohorts of 38 and 47 Masters students, specialising in either International Shipping or International Logistics.

It is conceivable that undergraduate responses may have been tainted by non-commitment bias (Evans, 1989), in that the choices and actions that they might make at some future date might not correspond with their current expressed preferences. Until their future actions and choices are revealed behaviourally, the extent of any such bias is impossible to determine.

For those registered students who had already made their decisions to undertake postgraduate study, reminiscence bias (Nisbett and Ross, 1980) could have been present in their responses. Their reminiscences of the issues, items and strength of perceptions which they reported as being significant post hoc their decisions to undertake study, may not have been the same as those prevailing at the time when the decisions were actually made. Hence, an attempt was also made to test the stability of retrospective perceptions, by repeating the instrument with one cohort of students. All 47 members in one Masters group were assessed initially, using a first survey in January 1998, which was the first combined meeting in Plymouth following either exemptions from Period One of the course, or completion of it at institutions abroad. The second survey was in March, at the end of their combined studies. Reminiscence data was the only source available post hoc the study decision and any tainting due to changed student expectations or emotional states, or lack of personal direct access to the mental states that influenced their actual behaviour was unavoidable.

An attempt was also made to discover any deeper underlying factors that might be influencing the decision to study ISL more succinctly, in order to refine possible future surveys. As a data reduction exercise, principal components analysis was employed to identify the major dimensions of statistical divergence within each main issue, for the 1997 and January 1998 data sets. All the factors identified with eigenvalues exceeding unity were included, and their individual contribution to the percentage of total variation present in the data is shown below. Although this approach presented a useful data reduction technique, interpretation of the precise meaning of factor loadings may be problematic. Findings are reported for the combined groups of 1997 and January 1998 students, totalling 85 responses by Masters students, and 52 by Diploma students. However, principal components analysis was not attempted for the undergraduate groups, due to statistical problems in dealing with incomplete responses, where many items were recorded as zero. Such null responses reflected either a lack of knowledge or interest in postgraduate study in many undergraduates, when the surveys were conducted.

Sample Characteristics

Diverse samples included more females in the Diploma groups (27 per cent) than in the undergraduate and Masters groups (15 per cent), although the mean age of postgraduates surveyed was 4-5 years older than undergraduates. In terms of the nationality of the undergraduates surveyed, many were of British (73 per cent), or South European origin (13 per cent), but several failed to respond. This contrasts with Diploma groups, which included French (42 per cent), Southern European (35 per cent), North European (19 per cent) and a few British (4 per cent) students. Masters students were mainly drawn from North Europe (46 per cent), South Europe (21 per cent), Asia (12 per cent) and Britain (8 per cent) with the rest coming from all parts of the globe. In terms of relevant work experience, 73 per cent of undergraduates reported that they had gained none, falling to 38 per cent of Diploma students and only 13 per cent of Masters students, with the mean length of experience rising from 0.8 to 1.9 to 2.8 years respectively. Half of the Masters students who reported relevant work experience had mainly worked inland, with the rest split equally between ports and seagoing work, whilst two-thirds of the Diploma students who had worked had worked at sea, with a further 20 per cent in ports, and relatively few inland. Seventy per cent of undergraduate experience was in ports, with the rest split equally between seagoing and inland work.

Diachronal Perceptions of Influences on the Study Decision

The repeated instrument, administered blindly to 47 Masters students for immediate return, afforded no realistic scope for any conscious recall of earlier answers, generating two independent data sets. Mean scores are reported for the January (score A) and March surveys (score B) and a Wilcoxon Matched-Pairs Signed-Ranks test (Wilcoxon, 1945) was used to test whether the perceived importance of items had changed over this period. The significance of items that had changed is shown in Table 5.1.

Post hoc perceptions of Masters level students were relatively stable, with no statistically significant variation on 44 of the 56 items tested. Increased mean scores on many items, even if not statistically significant, may reflect a heightening of student perceptions. An alternative explanation, of growing pride in their decision to study, in part helping to justify it retrospectively, had previously been observed in the undergraduate decision to study transport (Dinwoodie, 1996). Some items where perceptions had changed related to possible deterrents to undertaking postgraduate study, including the perceived importance of the need for operational experience, and a job offer affording good experience. Both of these effects were known to increase in significance towards graduation (Dinwoodie, in press), and fears of choosing the wrong course were also heightened. The 'retrospective justification' processes might account for the raised perceived importance of living in 'a nice place', and the potential for ratings of university facilities and courses, or poor perceived value for money, to deter prospective students. The perceived role in the decision to study of finding out about courses by chance, increased, but remained low overall, although the perceived role of a sufficient provision of course information heightened. An element of role reversal, where respondents now felt that they themselves could be considered to represent the experts may have focused their retrospective perceptions of the importance of bad reports in deterring them from enrolling.

Table 5.1 Items in the Masters level study decision: diachronal changes

Statements, mean scores and significant differences * = 95% level ** = 99% level	Mean score A	Change at mean score B	
I might have been put off study at postgraduate level because:			
I needed operational experience to improve my understanding.	0.24	+0.25	**
I had to be sure it was the right course: I could not repeat it.	0.60	+0.36	*
If I was offered a job giving good experience.	0.52	+0.27	*
I found out about these courses by chance.	0.28	+0.21	*
In the city of my university I would live in a nice place.	0.47	+0.27	*
I might have been put off study at my university if:			
It gave me too little information.	0.91	+0.35	*
It rated below other universities on its course quality, library, etc.	1.00	+0.38	*
It was not the best school which I could afford.	0.66	+0.30	*
Students had given me bad reports about it.	0.89	+0.54	**

Source: the author

In order to verify the temporal stability of the mean ratings of items by individuals between January and March, Spearman's non-parametric rho correlation coefficients were also computed for each item. The computed statistic was significant, or highly significant, on all items in the survey, so much so that results are not reported in detail. For the comparative distance measures, F3, F4, F5, F6, F7 and F8, 0.54 < rho < 0.86, all significant at the 99.9 per cent level. For the raw distance formula, F1, a rho value of 0.48, was significant at the same level. On all other items, rho was significant at least at the 95 per cent level.

Issues in the Overall Decision to Study Shipping or Logistics

Surveys were conducted of the perceptions of final year undergraduates, postgraduate Diploma level students, and students enrolled on Masters courses, with mean scores showing the importance of each issue and item in the study decision. In some tables, ranks of mean scores are shown. A non-parametric one way analysis of variance Kruskal-Wallis test of the differences between the measures of central tendency (Kruskal and Wallis, 1952) for the three distributions was used (shown KW), with statistically significant differences shown respectively at the 95 per cent, 99 per cent and 99.9 per cent level, as *, ** and ***. Similar conventions were used to denote the statistical significance of Mann Witney U tests of the differences in measures of the central tendency of distributions (Mann and Witney, 1947) between pair-wise samples of subgroups. These tests are reported for comparisons between Masters and Diploma students (Ms/Dp), Masters and Undergraduates (Ms/Ug) and Diploma and Undergraduates (Dp/Ug).

Principal components analysis of the broad issues which made postgraduate study in ISL attractive was not attempted, as it would have generated few useful results, given that each of the nine issues identified had already emerged from focus groups. In an analysis of the simple mean ratings and rankings of issues by subgroup, several key findings were apparent. Firstly, undergraduates registered much lower mean scores overall (Table 5.2) with no issue exceeding unity. Diploma students recorded four such scores, and Masters students seven. For undergraduates, the study decision was less immediate, and being one of only several course exit options open to them, was viewed as less important overall. For similar reasons, items that made postgraduate study of ISL attractive formed the most important issue in the study decision for both registered Masters students (mean score 1.68) and Diploma students (mean score 1.35), but ranked eighth for undergraduates. For undergraduates, family and friends and how to find out about courses predominated (both scoring 0.92). The former was a minor concern for postgraduates, but the latter interested all groups. Finally, the reasons that made study in the UK attractive concerned postgraduates but not undergraduates, who were already close to completing such a course.

Table 5.2 Rankings of issues in the study decision by subgroup

* Indicates a mean score exceeding 1	Ranking of issues		
Issues	MSc	Dip	Ug
Family or friends.	6= *	9	1=
Reasons which made studying logistics / shipping at postgraduate level attractive.	1 *	1= *	8
Reasons which might have put you off studying logistics / shipping.	9	5=	3
How did you find out about courses?	3 *	4 *	1=
What made studying logistics / shipping in the UK attractive?	2 *	1= *	5=
What reasons made studying logistics / shipping at your university attractive?	4 *	3 *	5=
What might have put you off studying logistics / shipping at your university?	6= *	5=	7
Why did you not study elsewhere?	5 *	8	9
Barriers which might have put you off studying at your university.	8	7	4
Items within barriers	Mean scores		
Money was an issue for me.	0.82	0.48	1.32*
The quality of life / my accommodation...	0.93	0.85	0.85
Teaching methods on the course.	1.02*	0.69	0.98

Source: the author

Reasons that might have deterred students or formed barriers to study, were given low rankings by registered postgraduates, but were more highly rated by undergraduates. In particular ex ante concerns over funding received the highest scores in the whole study for undergraduates. For postgraduates, this reflects either a reluctance to remember the downside ex post, or a failure to consider it at all. This could indicate either immaturity or a limited commitment to the decision that may signal a need for counselling or careers guidance intervention (Kidd and Killeen, 1992).

The Role of Family and Friends in the Study Decision

It has long been known (Board of Trade, 1970) that substantial proportions of seafarers hail from families with a seafaring tradition, inferring in turn that similar influences probably operate on students of international shipping. Similar trends have been observed on logistics and transport courses, where in one survey, some 45 per cent of undergraduates enrolled on a specialist undergraduate course in transport, claimed to know 'significant others' in the industry, two-thirds of whom were relatives (Dinwoodie, 1996, p.48). In the light of this, and also evidence from the focus groups which highlighted specific items relating to the role of family and friends in the study decision, perceptions of these items are discussed here.

The issue relating to the role of family and friends overall in the decision to study accounted for statistically significant variations (Table 5.3) between different groups. In particular, the Diploma groups were more independent of them in terms of funding or advice, and also felt that fewer family circumstances had been relevant to their decision to study. These views reflect the experiences of students in the international Diploma group, many of whom had worked extensively away from home, possibly at sea, and also European Diploma students, some of whom had chosen to study abroad because 'it made life more interesting'. Talking to family and friends had influenced Masters level students statistically significantly more than both other groups, and they also considered their family circumstances to be more relevant. This finding may be explained by their personal commitments to the needs of family businesses, which in some cases were funding their studies. Taken together, these groups represent different market segments, influenced by different considerations, which potential course providers need to be aware of.

Amongst the Masters level students, a principal components analysis of the influence of items in the family and friends issue on study decisions produced positive loadings against Factor 1 on all items. This represented a broad 'contact' factor loading highly against those students who perceived that their family circumstances, or who had been introduced to the industry by their family and friends, were important elements in their study decision.

Table 5.3 Influences of family or friends on the decision to study

Item scores and significant differences between subgroups	Mean scores			KW			
	Ms	Dp	Ug		Ms /Dp	Ms /Ug	Dp /Ug
This factor was an issue for me.	1.07	0.58	0.92	**	***		*
Family circumstances are relevant.	1.02	0.56	0.77	*	**		
They introduced me into the industry.	0.48	0.62	0.32				
I talked to them.	1.28	0.69	0.83	***	***	**	
They influenced me because they are funding my studies.	0.66	0.29	0.57	*	*		

How did your family or friends influence your decision to study? How important were they?

Factor scores in a principal components analysis of Masters replies.	Factor 1	Factor 2
% of variation explained by this factor:	46.9	25.9
My family circumstances are relevant.	0.834	-0.275
They introduced me into the industry.	0.715	-0.547
I talked to them.	0.650	0.422
They influenced me because they are funding my studies.	0.498	0.695

Principal components analysis of Diploma students.	Factor 1
% of variation explained by this factor:	66.3
My family circumstances are relevant.	0.860
They introduced me into the industry.	0.712
I talked to them.	0.791
They influenced me because they are funding my studies.	0.882

Source: the author

A secondary factor loaded positively on funding issues and negatively on being introduced to the industry by them. For Diploma students, the effects of using data reduction techniques were to produce a single factor, with positive loadings on all items. This factor represented a broad 'family and friends influence' perhaps best represented in cases where students were being funded by them or where family circumstances were perceived to be important considerations in their study decision.

Why Study Logistics or Shipping at Postgraduate Level?

Highly significant differences between the ratings of different groups on all items forming reasons to study at postgraduate level were apparent (Table 5.4). On all items except a desire to specialise and work in shipping, which were felt to be vital to students in the Diploma group but only moderately so in the Masters group which also included logistics specialists, ratings between the postgraduate groups were similar, but differed from undergraduate views. Issues of returning to study and changing career were important to Diploma students although scores still exceeded 1.2 despite a low ranking, with a strong desire to specialise in shipping and learn new things. For Masters students, a mix of knowledge and career issues predominated, but with employment opportunities forming important secondary concerns for all the groups. Taken overall, knowledge items on this issue recorded the highest mean scores found in the whole study (2.14), as might be expected, in a survey of the reasons for embarking on a course.

When analysed using principal components analysis, the reasons why Masters students undertook postgraduate study included a 'career change' factor loading positively on items relating to enacting long term career plans, career change, and broadening opportunities. A second factor related to a desire to return to study following work experience, and a third factor related to a desire to acquire new knowledge.

For Diploma students, a factor defining the search for 'new opportunities' was important in their decision to study shipping or logistics, loading positively on a desire to enact long term career plans, to return to study following work experience, and to broaden their opportunities. A second factor related to a desire to specialise and work in shipping, and a third factor to a desire to change career and go ashore, although loading negatively on a desire to broaden knowledge and learn new things. The interpretation of factors differs between Masters and Diploma groups on this issue.

Table 5.4 Attractions of postgraduate study of logistics or shipping

* (**) indicates a mean score exceeding 1 (2)	Ms	Dp	Ug
This factor was an issue for me (mean score)	1.68	1.35	0.75
I wanted:	Rank of issue		
to broaden my knowledge / learn new things.	1 **	2 **	3
to enact my long term career plans.	2 **	3 *	1
a change of career / to go ashore.	6 *	6 *	6
to broaden my opportunities / be sure to find a job.	3 *	4 *	2
to specialise in shipping as I wanted to work there.	4 *	1 **	4
more study, following my work experience.	5 *	5 *	5
Factor scores in a principal components analysis of Masters replies	Factor 1	Factor 2	Factor 3
% of variation explained by this factor:	28.2	19.7	16.9
I wanted:			
to broaden my knowledge / learn new things.	0.236	-0.123	0.916
to enact my long term career plans.	0.657	0.003	0.153
a change of career / to go ashore.	0.649	0.457	-0.274
to broaden my opportunities / find a job.	0.695	-0.316	-0.192
to specialise in shipping to work there.	0.537	-0.254	-0.044
more study, following my work experience.	0.098	0.889	0.187
Diploma replies			
% of variation explained by this factor:	28.1	20.0	16.7
I wanted:			
to broaden my knowledge / learn new things.	0.507	0.441	-0.607
to enact my long term career plans.	0.618	-0.299	-0.145
a change of career / to go ashore.	0.535	-0.455	0.544
to broaden my opportunities / find a job.	0.583	0.418	0.259
to specialise in shipping to work there.	0.248	0.652	0.430
more study, following my work experience.	0.602	-0.334	-0.251

Source: the author

Deterrents to the Study of Logistics or Shipping

The perceptions of reasons that might have deterred students from postgraduate study were not well developed in comparison to the positive attractions of study, for any group other than the undergraduates (Table 5.5). Shortages of funding for studies, the prospects of job offers affording good industrial experience, and a need to be sure that a course was the right one, all heightened undergraduate concerns. The extent of this effect was that the importance they attached to these items was statistically similar to that of postgraduates. However, items relating to waiting while they worked to accrue funds for study, and assurances that they were selecting the right course, concerned Diploma students more than Masters students. The only statistically significant source of variation between the groups occurred where Masters students felt that operational experience had little importance in their study decision, unlike some of the other groups. These findings imply that undergraduates, who might be considering the decision to embark on postgraduate study ex ante any commitment, have a stronger perception of the problems associated with it than those viewing it ex post. Marketing of postgraduate courses aimed at tempting mature returners from industry, should emphasise how apposite such courses are for particular groups of targeted individuals. By contrast, marketing aimed at undergraduates who might be considering further study might usefully extol the scope for funding their studies.

Data reduction using principal components analysis produced a first factor statistically influencing why Masters students might have been deterred from undertaking postgraduate study of shipping and logistics defining situations where an element of caution in the decision was important. Loadings on an inability to repeat the experience, possibilities of offers of work experience and work pressures, were important considerations in this factor. A second factor loaded positively on the material costs of returning to study, and a need to gain operational experience, and negatively on work pressures, defining a more material variable.

Table 5.5 Deterrents to the study of logistics or shipping

Item scores and significant differences between subgroups	Mean scores Ms	Dp	Ug	KW	Ms/ Dp	Ms/ Ug	Dp/ Ug
This was an issue for me.	0.88	0.79	0.88				
I did not yet have the money / was saving up, while working.	0.68	0.90	1.00				
I needed operational experience to improve my understanding.	0.24	0.70	0.70		***	***	***
Had I experienced work or family business pressures.	0.30	0.25	0.35				
I had to be sure it was the right course: I could not repeat it.	0.60	0.83	0.90				
If I was offered a job giving good experience.	0.50	0.80	0.98	*		*	
Returning to study after 10 years work would be hard.	0.34	0.42	0.47				

Factor (FA) scores in a principal components analysis of Masters replies	FA1	FA2
% of variation explained by this factor:	31.5	20.6
I did not yet have the money / was saving up, while at sea.	0.44	0.61
I needed operational experience to improve understanding.	0.23	0.74
Had I experienced work or family business pressures.	0.50	-0.53
I had to be sure it was the right course: I could not repeat it.	0.74	-0.17
If I was offered a job giving good experience.	0.77	-0.06
Returning to study after 10 years working would be hard.	0.50	0.14

Diploma replies	FA1	FA2
% of variation explained by this factor:	40.2	18.0
I did not yet have the money / was saving up while at sea.	0.36	0.83
I needed operational experience to improve understanding.	0.67	-0.22
Had I experienced work or family business pressures.	0.62	-0.33
I had to be sure it was the right course: I could not repeat.	0.63	0.35
If I was offered a job giving good experience.	0.69	-0.32
Returning to study after 10 years working would be hard.	0.75	0.08

Source: the author

For the Diploma students, a similar first factor related to situations where an element of caution in the decision was important also emerged. High positive loadings on items relating to fears of returning to studies, offers of work experience, a need for operational experience and family pressures defined this factor. Material issues defined the second factor that loaded positively, on a lack of funds and making sure that it was the right course, which could not be repeated. Taken together, these data reductions seem to indicate firstly an element of caution in the decision, and secondly an issue relating to material considerations.

How did Students Find Out About Courses?

Unlike on other issues where undergraduate scores were significantly lower than in other groups, there was relatively limited statistical variation between groups in the importance attached to sources of information about postgraduate courses (Tables 5.6 and 5.7). This observation furnishes evidence of exploration behaviour amongst applicants (Smart and Peterson, 1997), prior to enrolling. Many of the international Diploma students had made their decision to study in an industrial context, possibly when they were working at sea and as such, their decisions were influenced by oral communications made with industrial contacts. Course alumni were important influences on them. For undergraduates and Masters students, weighing up whether to undertake study whilst still in more academic surroundings, the credence which they placed on the views of their lecturers was substantial. More impersonal literary sources of information gained from reading magazines, books and brochures also formed an important element of the study decision for all groups. In terms of choosing suitable channels with which to market courses, sole-reliance on these literary sources of information is unlikely to succeed, as the evidence here suggests that a range of oral sources and information may be more significant. Influences include lecturers in an academic environment talking to students already enrolled on courses and contemplating further study, colleagues in an industrial setting, and the impressions of both alumni and prospective students.

Principal components analysis of the items influencing how Masters students had found out about postgraduate courses included a 'professional advice' variable. This was defined by a factor with negative loadings on discussions with lecturers and potential students, and positive loadings on work contacts and the British Council. A second dimension of divergence in the data loaded positively on chance and discussions with alumni. More remote influences such as literature, defined a third factor.

Table 5.6 Sources of information about courses (1)

Item scores and significant differences between subgroups	Mean scores Ms	Dp	Ug	KW	Ms/ Dp	Ms/ Ug	Dp/ Ug
This factor was an issue for me.	1.37	1.02	0.92	***	**	***	
I talked to a work friend / people in industry.	0.91	1.27	0.98				
My previous lecturers told me about it.	1.05	0.52	1.12	**	*		***
I contacted the British Council.	0.55	0.48	0.25				
I read magazines / books / brochures.	0.98	1.15	0.93				
By chance.	0.28	0.44	0.20				
I talked to others planning to study there.	0.61	0.65	0.77				
I talked to students / former students.	0.91	1.40	0.93	*	*		*

Source: the author

Data reduction techniques, used to generate factors which influenced how Diploma students had found out about a course, first defined a 'personal contact' component. This factor was defined by positive loadings on discussions with work contacts, lecturers and alumni, and negative loadings on more impersonal items including the British Council and literature sources. A second dimension of divergence in the data loaded positively on items involving discussions with students on the course, including both prospective students and alumni, but loading negatively on chance factors. A final factor loaded positively on chance factors. When comparing the analyses of Masters and Diploma level students, the interpretation of factors varied between the two groups, suggesting a need to tailor marketing campaigns to the specific needs of each group separately.

Table 5.7 Sources of information about courses (2)

How did you find out about courses in these areas? How important was each source?

Factor scores in a principal components analysis of Masters replies	Factor 1	Factor 2	Factor 3
% of variation explained by this factor:	25.6	20.8	15.6
I talked to a work friend / people in industry.	0.550	0.427	-0.126
My previous lecturers told me about it.	-0.786	-0.240	-0.017
I contacted the British Council.	0.673	-0.339	0.296
I read magazines / books / brochures.	0.137	-0.032	0.883
By chance.	0.180	0.578	0.045
I talked to others planning to study there.	-0.581	0.267	0.456
I talked to students / former students on the course.	-0.169	0.832	0.037
Diploma replies	**Factor 1**	**Factor 2**	**Factor 3**
% of variation explained by this factor:	25.4	23.5	15.8
I talked to a work friend / people in industry.	0.695	0.322	-0.370
My previous lecturers told me about it.	0.492	0.022	-0.433
I contacted the British Council.	-0.691	0.370	0.042
I read magazines / books / brochures.	-0.476	0.453	-0.202
By chance.	0.140	-0.649	0.544
I talked to others planning to study there.	0.090	0.751	0.431
I talked to students / former students on the course.	0.565	0.459	0.510

Source: the author

Why Study in the UK?

The problems of overseas students choosing to study in the UK are well documented (Harris, 1995), but enrolled postgraduates rated the advantages of this issue, and many items within it, more highly than undergraduates (Table 5.8). For the undergraduates surveyed, many of whom were close to attaining a UK qualification, there was relative indifference, but concerns to gain a qualification recognised worldwide, relevant to industry, and whilst abroad, were apparent. For Diploma students, mainly non-UK nationals, the desire to practise English, the language of shipping, and gain a UK qualification, recognised worldwide, and relevant to industry, were very important, and also less strongly, for Masters students. The desire to 'go abroad as it makes life more interesting' could have been satiated in many other countries elsewhere, but the supporting desire to study in English made study in England all the more attractive. A range of factors influenced other students. These included the length of time required to complete courses, a land with a long shipping tradition, or course availability and were secondary considerations that may have made the UK more attractive than the USA, Australasia or the Nordic lands.

In the data reduction exercise, the importance of the attraction of courses in the UK, in the study decision of Masters level students included a 'utilitarian' factor. This recorded high loadings on the international recognition afforded to British qualifications, and the lack of suitable courses or opportunities to speak English, the language of shipping, in their own country. A second factor loaded negatively on a desire to go abroad, and positively on an academic system relevant to industry and to study in a land with a long shipping tradition.

For Diploma students, the importance of the attraction of courses in the UK in the study decision also included a 'utilitarian' factor, with high loadings on the brevity of courses, the international recognition afforded to British qualifications, their industrial pertinence and the importance of the UK's shipping tradition. A second factor represented a desire to go abroad, for reasons of general interest, and to practise the English language, also loading negatively on a paucity of courses at home. Again, the components defined at Masters and Diploma levels do not appear to be directly interchangeable.

Table 5.8 Attractions of studying logistics or shipping in the UK

* (**) indicates a mean score exceeding 1 (2)	Ms		Dp		Ug
This factor was an issue. Item means score	1.49		1.35		0.80
I wanted:	Rank and significance				
to go abroad as it makes life more interesting.	3	*	3	*	2
a British Diploma or MSc, recognised worldwide.	1	*	1	**	1
an MSc: we don't have them in my country.	5		6		6
to practise English / the language of shipping.	2	*	2	*	5
a UK course which can be completed quickly.	6		7		7
an academic system more relevant to industry.	4		4	*	3
to study in a land with 500 years of shipping.	7		5	*	4

Factor (FA) scores in a principal components analysis of Masters replies	FA1	FA2
% of variation explained by this factor:	27.0	24.4
I wanted:		
to go abroad as it makes life more interesting.	0.451	-0.722
a British Diploma recognised around the world.	0.628	-0.209
an MSc: we don't have them in my country.	0.607	0.424
to practise English / the language of shipping.	0.594	-0.382
a UK course which can be completed quickly.	0.432	-0.236
an academic system more relevant to industry.	0.514	0.600
to study in a land with 500 years of shipping.	0.346	0.635

Diploma replies	FA1	FA2
% of variation explained by this factor:	35.9	21.6
I wanted:		
to go abroad as it makes life more interesting.	0.135	0.876
a British Diploma is recognised around the world.	0.648	0.208
an MSc: we don't have them in my country.	0.608	-0.457
to practise English / the language of shipping.	0.625	0.452
a UK course which can be completed quickly.	0.708	0.219
an academic system more relevant to industry.	0.612	-0.443
to study in a land with 500 years of shipping.	0.660	-0.215

Source: the author

Why Study at a Particular University?

The mean scores relating to the importance of reasons for choosing to study at a particular university recorded amongst undergraduates were statistically significantly lower than those for postgraduates (Table 5.9). The highest recorded scores related to a specialisation in shipping, perceived as being very important to Diploma students, and also to Masters students, although rather less so. Other items related specifically to particular courses or places, in addition to broader issues concerning the choice of their particular subject or country in which to study. The reasons that made a particular university attractive were less important than the choice of subject, country, or finding out about courses.

When considering the importance of the characteristics of the city in which the chosen university was located, and why students might be attracted to study in it, no outstandingly high ratings were recorded in any of the groups. It was apparent that inertia, associated with already having studied in the city, and a certain loyalty or pride in having taken their earlier decisions to study there, produced a relatively higher rating of these factors by undergraduates. For both Masters and Diploma level students, these items were rated lower than other items, emphasising the predominance of vocational development and course issues amongst their concerns. The item relating to knowledge of friends who had completed the course was rated the lowest of all items in this set, somewhat surprisingly, given the importance attached to oral sources of information that was noted earlier. Items relating to the reputation of the university were rated more highly. In particular, the perceived importance of improved employment prospects associated with having studied at a university which awarded qualifications that were widely recognised, was high. This item appealed to Diploma students even more strongly than to Masters students, and also formed the highest single undergraduate item score on this issue. For Diploma students, commendations by former lecturers, often overseas, were also important.

With regard to more general items relating to the choice of university, the offer of a place and the reputation of its lecturers were important, particularly to Diploma students, but also to Masters students. Undergraduates were attracted by few of these items, but were relatively more concerned that their prospective lecturers had a good reputation than by other items.

Table 5.9 Attractions of logistics / shipping at my university

Item scores and significant differences between subgroups	Mean scores Ms	Dp	Ug	KW	Ms /Dp	Ms /Ug	Dp /Ug
This factor was an issuc	1.18	1.15	0.80	**		**	**
My university's reputation:							
was commended by my lecturers.	0.58	0.90	0.60				
is worldwide, important when looking for a job.	1.04	1.25	0.75	*			**
My university's course:							
is the only one which offers an MSc.	0.38	0.46	0.40				
is the only one in international logistics.	0.49	1.04	0.40	***	***		***
specialises in shipping.	1.16	2.13	0.62	***	***	**	***
offers an MSc: the Diploma is an entry …	0.66	1.23	0.38	***	**		***
could give me remission from the MSc Stage 1.	0.44	1.06	0.37	**	***		**
In the city of my university:							
I have friends who have finished the course.	0.29	0.42	0.35				
I have already undertaken studies.	0.39	0.58	0.65				
I would live in a nice place / beautiful city.	0.47	0.54	0.62				
My university:							
has a good admissions administration.	0.52	0.79	0.47				
accepted me.	1.06	1.31	0.52	***		**	***
did not have bad oral reports …	0.58	0.77	0.42				
lecturers have a good reputation.	0.85	1.39	0.65	**	*		**

Source: the author

Perceptions of courses were an important element in the study decision. In relation to the course reputation, a specialisation in shipping was considered to be important for all groups. This was particularly so for the Diploma students, who were also attracted by the scope for progression to, or given exceptionally good results, partial exemption from, the Masters programme. Many individuals were anxious to advance their careers, expressing this in items relating to their particular interest in courses. The importance of selecting courses that they perceived to be unique, or that offered the opportunity to pursue a particular specialist discipline in depth were highlighted in this context.

Similar ratings of these issues were also apparent in the principal components analysis of items within this issue. At Masters level (Table 5.10), five factors defined the importance of the attractions of a particular university. In decreasing order of significance, these were interpreted as:

1. 'Reputation', loading on not having bad oral reports, the reputations of the course and its lecturers, and a good admissions policy.
2. Course uniqueness, with high loadings on commendations of lecturers and course identity.
3. 'Opportunities', including high positive loadings on the potential for Masters level study and negative loadings on 'acceptance'.
4. 'Inertia', with high loadings on personal or social links with alumni.
5. The importance of studying in 'a nice place'.

Table 5.10 Reasons why Masters students chose Plymouth

What reasons made the study of logistics or shipping at Plymouth attractive? How important was each?

Factor (FA) scores in a principal components analysis of Masters replies

% of variation explained by this factor:	FA1 26.8	FA2 13.7	FA3 9.7	FA4 8.4	FA5 7.2
Plymouth's reputation:					
was commended by my lecturers.	0.40	0.591	0.210	0.036	-0.217
is worldwide, important when looking for a job.	0.76	0.246	0.098	-0.068	-0.027
The Plymouth course:					
is the only one which offers an MSc.	0.484	0.218	0.287	-0.217	-0.441
is the only one in international logistics.	0.095	0.554	0.446	-0.307	0.220
specialises in shipping.	0.546	-0.453	-0.052	-0.302	-0.149
offers an MSc: the Diploma is an entry.	0.309	-0.601	0.565	0.150	0.138
could give me remission from the MSc Stage 1.	0.297	-0.697	0.384	0.027	-0.133
In the city of Plymouth:					
I have friends who have finished the course.	0.432	0.129	0.223	0.573	-0.247
I have already studied.	0.339	0.114	-0.212	0.689	0.163
I would live in a nice place / beautiful city.	0.463	0.036	0.194	0.016	0.668
Plymouth University:					
has a good admissions administration.	0.659	0.070	-0.089	-0.246	0.303
accepted me.	0.502	-0.236	-0.500	-0.205	-0.060
did not have bad oral reports.	0.705	0.075	-0.176	0.045	-0.155
lecturers have a good reputation.	0.763	-0.044	-0.368	0.036	0.078

Source: the author

At Diploma level (Table 5.11), some of the factors which had emerged at Masters level were still broadly recognisable, but in changed order of importance. In diminishing order of significance they included:

1. A factor relating to 'admissions' issues, loading highly on items which portrayed the Diploma as affording an entry to Masters courses, the importance of being accepted on the course, admissions policy, recognition given to prior study and subject uniqueness.
2. 'Reputation' including positive loadings on commendations of the course by lecturers, and a worldwide reputation associated with it.
3. A 'subject' factor, including high loadings on the importance of specialisation in shipping and being accepted on the course.
4. A further subject factor loading positively on a specialism in shipping and the uniqueness of a Masters level qualification, and negatively on prior study at the institution.
5. A factor loaded highly on the 'social' aspects, where social contacts with alumni were considered to be important in the decision to study.

In attempting to generalise in relation to what makes a particular institution attractive, it is apparent that the 'reputation' of the institution, its courses and its lecturers is important in all groups. Subject issues, where particular courses are perceived as 'unique' or specialist form a major attraction. Opportunities for personal advancement, represented at Diploma level by the possibility of academic progression to Masters courses, and at Masters level by enhanced employment prospects were coupled with the importance of being offered a place and hence the prospect of advancement. Social, inertial and locational concerns were apparently secondary considerations for most students.

Table 5.11 Reasons why Diploma students chose Plymouth

What made the study of logistics or shipping at Plymouth (or St Nazaire / Plymouth for Eurodip students) attractive? How important was each?

Factor (FA) scores in a principal components analysis of replies

	FA1	FA2	FA3	FA4	FA5
% of variation explained by this factor:	30.9	14.1	8.9	8.4	7.4
Plymouth's reputation:					
was commended by my lecturers.	-0.064	0.679	-0.212	0.121	0.174
is worldwide, important when looking for a job.	-0.028	0.623	0.377	0.127	-0.429
The Plymouth course:					
is the only one which offers an MSc.	0.444	-0.029	-0.432	0.492	-0.308
is the only one in international logistics.	0.599	-0.473	-0.295	0.098	-0.027
specialises in shipping.	0.400	-0.043	0.523	0.554	0.061
offers an MSc: the Diploma is an entry.	0.832	-0.148	-0.138	0.057	0.075
could give me remission from the MSc Stage 1.	0.656	0.170	-0.196	0.372	-0.052
In the city of Plymouth:					
I have friends who have finished the course.	0.334	0.205	0.078	0.212	0.747
I have already studied.	0.558	0.272	-0.356	-0.442	-0.059
I would live in a nice place / beautiful city.	0.272	-0.791	-0.090	-0.037	-0.018
Plymouth university:					
has a good admissions administration.	0.823	-0.107	0.022	-0.185	0.085
accepted me.	0.674	-0.161	0.470	-0.103	-0.261
did not have bad oral reports.	0.627	-0.031	0.279	-0.279	0.278
lecturers have a good reputation.	0.678	0.246	0.127	-0.289	-0.063

Source: the author

Deterrents to Study at a Particular University

Masters level students placed slightly greater importance on the reasons that might put them off studying at their university compared with other groups (Tables 5.12 and 5.13). They may simply have been more aware and articulate, or possibly, given that most were unlikely to proceed further with their academic careers, they may have felt more critical, unlike many Diploma students, still seeking to advance academically. The offer of a place to study at Masters level was particularly important to them, in contrast to Diploma students who were more concerned that potential employers would consider their qualifications to be sufficiently unique.

Table 5.12 Deterrents to study at my university (1)

Item mean scores and significant differences between subgroups	Means			KW		
	Ms	Dp	Ug	Ms /Dp	Ms /Ug	Dp /Ug
This factor was an issue	1.07	0.79	0.78			
If I had found that my University:						
gave me too little information.	0.91	0.94	0.85			
was costly in relation to other courses.	0.84	0.87	0.50		*	
rated below other universities on its course quality, library, etc.	1.00	1.06	0.78			
had not offered me a place at MSc level.	1.42	0.88	0.57	***	*	***
qualifications were not considered unique by employers.	0.81	1.15	0.45	**	*	***
considered my grades to be inadequate.	0.92	0.96	0.42	*	**	**
was not the best school which I could afford.	0.66	0.79	0.38			*
students had given me bad reports about it.	0.89	0.94	0.43	**	**	*

Source: the author

For both groups, the ratings of their university in terms of its course quality and library were important, as was the need to be given sufficient information, an item also of importance to undergraduates.

Table 5.13 Deterrents to study at my university (2)

What reasons might have put you off the study of logistics or shipping at Plymouth? How important was each?

Factor (FA) scores in a principal components analysis of replies

Masters students	Factor 1
% of variation explained by this factor:	62.4
If I had found that Plymouth University:	
gave me too little information.	0.732
was costly in relation to other courses.	0.818
rated below other universities on its course quality, library.	0.885
had not offered me a place at MSc level.	0.789
qualifications were not considered unique by employers.	0.839
considered my grades to be inadequate.	0.807
was not the best school which I could afford.	0.740
students had given me bad reports about it.	0.694

Diploma students	Factor 1
% of variation explained by this factor:	61.6
If I had found that Plymouth University:	
gave me too little information.	0.753
was costly in relation to other courses.	0.698
rated below other universities on its course quality, library.	0.807
had not offered me a place at MSc level.	0.741
qualifications were not considered unique by employers.	0.806
considered my grades to be inadequate.	0.837
was not the best school which I could afford.	0.797
students had given me bad reports about it.	0.827

Source: the author

Principal components analysis revealed one main factor influencing the importance of reasons which might have deterred Masters students from study at Plymouth, defining a variable relating to how 'reasonable' the university was perceived to be. High positive loadings were recorded on the importance of value for money, the offer of a place on a Masters level course, reasonable resources ratings, and recognition of awards by employers. However, no single item could fully represent its rating.

Similarly, at Diploma level, a single main factor influenced the importance of reasons that might have deterred students from study at a particular institution. This again related to how 'reasonable' the university was perceived to be. The factor was defined by high positive loadings recorded on the importance of the offer of a place on a Masters level course, sufficient information, reasonable resources ratings, recognition of awards by employers and generally good value for money, with no adverse reports from students.

Taken overall, the less 'unreasonable' a university was perceived to be, then the less likely it would be to deter applicants. Masters students were primarily concerned with the offer of a place, Diploma students with the currency of their qualifications, and undergraduates with receiving sufficient information. All were concerned about the ratings of facilities, course quality and the like.

Why Not Study at Other Universities?

A similar set of considerations concerned students in deciding whether to study at other universities (Table 5.14). Although this issue was statistically more significant to Masters students, relatively low scores overall reflect a reluctance to consider alternative courses by many individuals, producing zero ratings on some items. There may also have been relative indifference amongst students, indicative of a competitive market for Masters courses. Noteworthy considerations included issues of subject matter, course emphasis, the time required to complete the course, and the total cost to the student. For Diploma students, the items related to how unique the course was considered to be, with the highest ratings recorded against the subject matter item, although cost was also a factor. For undergraduates, the same two items predominated.

Table 5.14 Reasons why students did not study elsewhere

Item scores and significant differences between subgroups	Means			KW			
	Ms	Dp	Ug		Ms /Dp	Ms /Ug	Dp /Ug
This factor was an issue	1.08	0.71	0.65	**	*	**	
Elsewhere:							
it was a different subject.	0.89	1.00	0.70				
I would have had to repeat material.	0.54	0.58	0.55				
the course emphasis was wrong for me.	0.81	0.54	0.62				
it was very expensive, and not necessarily better.	0.86	0.79	0.67				
the course took longer.	0.87	0.62	0.52				

Factor scores (FA) in a principal components analysis of replies

Masters students	FA1	FA2
% of variation explained by this factor:	48.9	20.3
Elsewhere:		
it was a totally different subject.	0.746	-0.242
I would have had to repeat material I had already studied.	0.731	-0.252
the course emphasis was wrong for me.	0.767	-0.462
it was very expensive, and not necessarily better.	0.636	0.557
the course took longer to complete.	0.601	0.606

Diploma students	Factor 1
% of variation explained by this factor:	45.2
Elsewhere:	
it was a totally different subject.	0.656
I would have had to repeat material I had already studied.	0.706
the course emphasis was wrong for me.	0.519
it was very expensive, and not necessarily better.	0.735
the course took longer to complete.	0.723

Source: the author

Principal components analysis of the factors which summarised the importance of reasons why Masters students chose not to study elsewhere in the overall study decision, revealed two main dimensions of divergence in the data. The major factor which was identified loaded highly on items relating to 'it was not what I wanted', with subject content, repetition of material, and a 'wrong emphasis' being important considerations. A less significant factor loaded heavily on the time and cost that was required to complete a course.

For Diploma students, a single factor emerged from the principal components analysis which summarised the importance of reasons for not choosing to study elsewhere in the overall study decision. This factor loaded heavily on the cost or time required to complete a course, an inappropriate subject content or emphasis, or a course on which repetition of material would be experienced. This factor 'explained' a relatively low percentage of the data, and can not be summarised easily by a single item in the list. A resource issue, relating to the time and cost that was required to complete a course, was apparent at both Masters and Diploma level, which had deterred students from pursuing their studies elsewhere. However, a more fundamental issue relating to individual choice and how individuals perceived that their particular needs were most likely to be met by a particular course, formed the main consideration for Masters students in this study.

Barriers to Postgraduate Study

The importance of barriers which might have put students off studying international shipping and logistics at their university were analysed in relation to pecuniary, personal and pedagogic issues which were raised in the earlier focus groups. Although not highly rated for any group (Table 5.15), scores for the Masters group statistically exceeded those for the Diploma group, with undergraduate scores also relatively high, as expected. Pecuniary considerations predominated for undergraduates, but were less of a problem for Diploma students who had spent time working in industry to save up for their studies. Corporate or parental funding and encouragement, particularly important to many Masters students, were also important to other groups. The quality of life was a consideration for all groups, although not of overriding importance. Pedagogic issues did potentially present important barriers to undertaking study, for all groups, but particularly the Diploma students, keen to visit companies and form professional links. The desire by students who sought to understand material,

Table 5.15 Barriers to study at my university (1)

Mean scores and statistically significant subgroup differences	Means			KW			
	Ms	Dp	Ug		Ms /Dp	Ms /Ug	Dp /Ug
This was an issue for me	0.98	0.73	0.87		*		
Money:							
This was an issue for me.	0.82	0.48	1.32	**	*	*	***
I was willing to pay a bit more for what I wanted.	0.48	0.77	0.70				
The total cost was less, when I considered the time involved.	0.42	0.42	0.63				
My company or parents partly funded and encouraged me.	1.05	0.75	0.70				
The quality of life in the city, my accommodation.	0.93	0.85	0.85				
Teaching methods on the course:							
This was an issue.	1.02	0.69	0.98		*		
My first contacts with staff made me welcome.	1.27	1.21	0.85	*		*	
I expected to understand, not merely to analyse statistics.	1.27	1.17	1.07				
My practical experience was undervalued in entry requirements.	0.47	0.42	0.62				
I expected visits to companies and form professional links.	1.08	1.35	1.17				

Source: the author

rather than merely to analyse statistics, implies a hunger for deep learning (e.g. Entwistle et al, 1992), which course planners would do well to ponder. Finally, the need to feel welcome was important.

When the data for Masters students was analysed using principal components analysis (Table 5.16), the importance of barriers which might have put students off studying shipping or logistics at their university were analysed in relation to material and pedagogic issues. In terms of monetary considerations, the first factor identified related to those students who were willing to pay for what they wanted and also keen to use their time to best effect. A second loaded highly on those students who were receiving parental funding. In terms of pedagogy, one single factor loaded positively on all items, including the importance of being made to feel welcome on arrival, and expectations of a desire to 'understand' material, and make industrial contacts whilst on the course.

Diploma students generated a single main factor in each of the principal component analyses relating to barriers which might have put students off studying shipping or logistics at their university, analysed in relation to material and pedagogic issues. In terms of monetary considerations, the main factor loaded highly on all items, but especially those willing to pay for what they wanted. In terms of pedagogy, the factor related to the importance of industrial links that could be established whilst on the course, the importance of being made to feel welcome on arrival, and expectations of a desire to 'understand' material.

Although not identical, the factors relating to a willingness to pay for what they wanted probably defined the main potential material barrier to undertaking study. This related to an expectation of students to understand what was going on in classes, supported with professional links, and an initial feeling of being welcomed by the institution.

Table 5.16 Barriers to study at my university (2)

How important is each of the following barriers that might have put you off the study of logistics or shipping at Plymouth?

Factor (FA) scores in a principal components analysis of replies by Masters students	Factor 1	Factor 2
% of variation explained by this factor	49.6	25.2
Money:		
I was willing to pay a bit more for what I wanted.	0.760	-0.437
The total cost was less, when I considered the time involved.	0.728	-0.507
My company or parents partly funded and encouraged me.	0.656	0.539

Teaching methods on the course:	Factor 1
% of variation explained by this factor:	55.9
My first contacts with staff made me feel welcome.	0.719
I expected to understand, not merely to analyse statistics.	0.877
My practical experience was undervalued in entry requirements.	0.563
I expected visits to companies and professional links.	0.794

Diploma students. Factor (FA)	Factor 1
% of variation explained by this factor:	69.3
Money:	
I was willing to pay a bit more for what I wanted.	0.902
The total cost was less, when I considered the time involved.	0.760
My company or parents partly funded and encouraged me.	0.748

Teaching methods on the course:	Factor 1
% of variation explained by this factor:	60.4
My first contacts with staff made me feel welcome.	0.874
I expected to understand, not merely to analyse statistics.	0.860
My practical experience was undervalued in entry requirements.	0.485
I expected visits to companies and professional links.	0.825

Source: the author

Summary

Surveys of the importance attached to most of the issues and items found to be important in the decision to undertake advanced study in ISL revealed that undergraduates generally rated them less highly than Masters and Diploma students. As hypothesised, ex post ratings statistically significantly exceeded ex ante perceptions of the study decision in relation to the importance of the positive attractions of the subject area, the country of study, and the particular university course. Far less variation was apparent in relation to the perceived importance of reasons for not studying a particular subject, or at a particular university, or the barriers to study. As noted in earlier work (Dinwoodie, 1996), before a decision to study has been made, prospective applicants are very wary of the potential pitfalls associated with the decision, but once it has been made, a retrospective justification is apparent. Eventually, not only are the perceived attractions of the course of action inflated, but pride in having taken a decision can even lead to an inertia against moving on at the end of the period of study.

There was little evidence of significant shifts in the perceived importance of items between independent surveys of reminiscences of the study decision taken at between three and six months after enrolment. Non-parametric statistical comparisons of the responses of one cohort of 47 Masters students revealed no significant differences on 44 out of 56 items tested. Exceptions may provide further evidence of retrospective justification, where the perceived importance of being in a nice place, and potential for poor facilities, information or value for money or bad oral reports to act as deterrents, heightened. Increased lapsed time post hoc the study decision also heightened perceptions of the potential of good job offers, the need for work experience, and fears of choosing the wrong course to act as deterrents to study. Chance was also perceived to have played a greater role in finding out about courses.

Subject content presented the main attraction of courses in ISL for Diploma students, but opportunities to acquire new knowledge and broaden employment opportunities were of greater concern to Masters students. Deterrents to study were often pecuniary, involving either experienced individuals waiting for several years while working to save up to study, or competing offers of employment for undergraduates, observed to be displaying caution in choosing the 'right' course.

Word of mouth was an important source of course information, where talking to industrial contacts in the workplace had influenced mature students, and also where lecturers' recommendations had influenced undergraduates and Masters students. Discussions with alumni, enrolled students and applicants also influenced choices, in addition to written materials.

The decision to study in the UK depended partly on the perceived recognition afforded to qualifications, partly to study in English, but for some students it was simply 'more interesting' to study overseas. Students were attracted to particular institutions because of their subject specialisms, but the reputation of the course and its lecturers, the offer of a place, and the prospects of proceeding to higher awards were also important. Students may have been deterred by poor ratings of institutions, inadequate information or bad reports. Perceived barriers to undertaking study at their university were found to form where students might feel unwelcome, fail to form professional links on their courses, or pedagogically, fail to 'understand', and not acquire deep learning.

In order to explain further any systematic similarities or differences between the decisions of particular individuals, or groups of individuals, it is now necessary to analyse links between the issues and items identified in making study decisions. In particular, the decision will be represented as a structured set of cognitive experiences, which can be presented in map form for each individual. Differences between these maps can then be quantified and analysed statistically.

6 Mapping and Comparing Postgraduates' Study Decisions

Introduction

Decisions to study shipping and logistics are not based on perceptions of a few issues standing in isolation, but rather the whole decision situation, influenced by the prior experiences of an individual. This chapter aims to analyse links which individuals make between the hereto isolated issues and items identified as influencing their decisions, and how these links vary between individuals and groups. In doing so, the technique of cognitive mapping is used first to represent, and later to compare individual decisions. This offers a powerful means of structuring and analysing complex decisions, unlikely to be familiar to most managers in shipping and logistics. To assist the layman, the discussion first presents approaches to cognitive mapping as a means of representing knowledge, before building a cognitive map of a decision situation relating to where to run a short course. The scope for using computers in cognitive mapping is then reviewed, before outlining some of the attempts to quantify differences between maps. Next, the methodology used to transform the instrument developed from the focus groups investigating the decision to study International Shipping and Logistics into a cognitive map is described. Examples of the exploratory results obtained from comparing student maps, and the student reactions towards using technological aids in re-running these experiments are discussed. The chapter ends by analysing some of the factors that were found to influence differences between student maps.

Building a Cognitive Map of Where to Attend a Short Course

Some Approaches to Cognitive Mapping

Before describing the practical steps in building a cognitive map of the process of deciding where to attend a short course, some approaches to mapping will be discussed. Many of the concepts of cognitive mapping were predicated on Kelly's personal construct theory (Kelly, 1955), concerned with investigating the ways in which individuals attempt to understand their world. In essence, the aim of the approach was to assist decision-makers to make better predictions about the impacts of alternative courses of action in future situations that may confront them. Although a family of useful business techniques have been identified (e.g. Spicer, 1998), they are all based largely on statements by individuals which are reliant on their experience of phenomena in the real world. The approach assumes that if an individual can predict social situations correctly, they will be better equipped to decide how to act in order to control the expected outcomes, and achieve their preferred goals. The technique can be used to analyse verbal or documentary accounts of problem contexts.

Bogoun (1983) hypothesised that individuals would express themselves using words based on 'raw experience, ...energy flowing through the skin ...upon which we erect our perceptions, knowledge and epistemological systems' (p.173). Two levels of schema, acting as pattern recognition devices were postulated, including at one level, schemas which transform raw experience, in its entirety, into knowledge, and further schemas, which organise and retain knowledge. In conceiving experience as a whole, a pattern becomes associated with a concept, so that if a name tag attached to a face or melody is experienced, this tags a concept, in turn tagging a pattern represented by an abstract, imageless and wordless element of thought.

The ability to reconstruct and explore a person's concept structure rests on the assumption that words can tag the concepts in the structure, so that a cognitive map is defined where the territory of verbal concepts have been recorded on paper. Non-verbal concepts only become expressible, when a socialising experience, results in labels that can be understood by at least one other person can be attached to them. Bogoun (1983) devised a 'Self-Q' interview technique to explore a person's concept structure, recorded in a concept map. Schema-concept pairs represent non-verbal patterns of relations among concepts, whose meaning will involve spatial representation of the patterns, in a cognitive map. These could vary from a cause map of causality relations, to a mapping of social territories, but meaning requires at least two sensations to be mapped.

The cognitive map includes both the individual's enacted experiences, represented in their schema structure, their enacted perceptions of self which they have constructed, and the relationships between the two. These schemae may emanate from reflexes or the first postnatal movements, and have evolved ever since (Piaget, 1967). Eventually, by projecting complex schemas, a social externality can develop, but unless an object is internalised, it does not exist. In so far as that a cause map is a map of a person's motivation structure, their activities only make sense in the context of the meaning they attach to them. Only when a non-threatening interviewer is able to access these maps can they begin to understand these social territories. Bogoun (1983) used a four stage interviewing technique, collecting concepts initially, later verifying and sorting them, then obtaining causality relations, before finally verifying that they made sense to the respondents themselves.

In building cognitive maps, an account of a problem situation is typically broken into phrases of 10-12 words, which represent distinct concepts. Contrasting phrases or 'poles' are sought, which enable concepts with a distinct meaning to be identified, and eventually these are ordered hierarchically to retain meaning through their context. 'Goals' are the most superordinate concepts, and are often linked to strategic directions in thinking (Ackermann et al, 1991). There is no underlying 'right' answer, given the ontological privilege noted above, but the mapping process provides a useful means of representing an individual's perception of a complex situation, preserving their own words where possible. A set of subordinate options emerges, linked hierarchically by arrows to their higher order outcomes, until eventually a more complete representation of the problem emerges.

The Process of Building A Cognitive Map of Where to Study

One approach to building a cognitive map of a problem situation involves the ordering and development of statements made by a problem owner in an initial general statement of a problem. If necessary, the approach may be attempted individually, commencing with a simple statement in which the problem owner rambles around the problem situation. Alternatively and preferably, the approach could be a social one, in which a consensus view of a problem is developed, possibly with the assistance of an independent researcher who merely records and attempts a trial ordering of concepts as they are expressed by the group members. In the case where one individual develops the map alone, the initial problem description may be recorded in an interview. This may be transcribed later, or written down as a personal and unpremeditated statement of a problem context, which expresses the individual's own understanding of the situation in their own words. The statement of the problem, is an attempt to record the situation as perceived by the problem owner at the time when the statement was made, as in the example presented below. It can make no claim to represent a consensus view, or even to be comparable prima facie in any way, with statements made by other problem owners faced with similar situations. In order to effect successful execution of the mapping procedures, it is important that a competent researcher, who has received suitable training, is involved. If the problem owner is not familiar with approaches to building cognitive maps, help should be sought, from a trained researcher.

In the example shown, a regional manager was considering whether to support employees in attending short courses of study, designed for practitioners. His problem related to the geographical location of the course which employees should attend. A simple initial statement of part of the problem situation is repeated verbatim.

> We weren't sure whether to attend these short courses in Plymouth or London. We could attend the course in London, but that would incur higher accommodation costs, as we don't live there, or higher travelling costs for staff, but might make a wider range of options available. We could attend in Plymouth, which might give us local expertise to tap, and would certainly be a much pleasanter place to be in early September. For local staff or anybody who has studied at Plymouth before, it might be more convenient to be in a more intimate environment. We could try a split between Plymouth and London, but that might fail to fully satisfy all staff and would still incur the administrative burden of finding accommodation and paying fees at both sites. Funding support for employees' attendance may be in jeopardy, if we send them outside of Plymouth. However, experience has shown that if we do not attend our business will suffer, but if we get it right, this could be the way forward for a range of employees on different types of courses...

The technique of cognitive mapping is used to organise, analyse and assist in understanding statements of concern about a problem situation, and when working with written statements, it is important to read the whole statement first to get a 'feel' for the problem. Once an individual begins to understand their world, George Kelly's theory of personal constructs postulated that they are in a position to decide how to attempt to change it, by attempting to take some kind of action or intervention in it. By implication, if an individual can make sense of their world, then, assuming that it still behaves the same way in future as it has done in their past experience, they can begin to understand what they might be able to do in order to achieve their goals.

Three processes can be identified in one approach to applying this technique, based on the mnemonic COG. Once relevant concepts (C) have been identified in a problem statement, a process of making comparisons by searching for opposites (O) ensues, before these are eventually hierarchically ordered into a graph (G). The processes involved are:

1. Ordering of thoughts into identifiable concepts. Using the words of the original statement, break it down into phrases of up to about ten words in length, which can later be re-ordered graphically to re-state the problem.
2. Looking for meaningful comparisons, or opposites linking two phrases. Constructs take on meaning through the process of making comparisons.
3. A graph can be traced, linking the ordering of phrases, relative to each other. In the ordering of goals and actions, phrases take on meaning in terms of their relative position in the hierarchy of concepts.

Process One: Order Thoughts into Identifiable Concepts

Three stages are involved in identifying concepts:

1a) Break the original statement into phrases of about ten words.

1b) Graphically order the phrases, with goals at the top, supported by strategies for achieving them, and any other optional courses of action lower down. Goals, at the top of the hierarchy, are not open to discussion or compromise, and may be apparent by such traits as voice modulation in an oral discussion. They may be placed well down a written text.

1c) Supporting strategies indicate the possible direction of courses of action. As actions, they will not be short term, low cost, reversible, or easily achieved; they may be supported by other options.

Eventually, detailed concepts will be identified which may lead to others becoming achievable (e.g. 'obtain cash' might lead to 'go to the bank'), indicating the options. In the example, open and closing brackets ( ) indicate the start and end of phrases, and possible goals are underlined.

> We weren't sure whether to (attend) (these short courses) (in Plymouth or London). We could (attend the course in London), but that would incur (higher accommodation costs as we don't live there), or (higher travelling costs for staff), but might make a (wider range of options available). We could (attend in

> Plymouth), which might give us (local expertise to tap), and would certainly be a (much pleasanter place to be) in early September. For (local staff) or (anybody who has studied at Plymouth before), it might be (more convenient) to be in a (more intimate environment). We could try a (split between Plymouth and London), but that might fail to (fully satisfy all staff) and would still (incur the administrative burden of finding accommodation and paying fees) at both sites. (Funding support for employees' attendance) may be in jeopardy, if we send them (outside of Plymouth). However, (experience has shown) that if we (do not attend) our (business will suffer) but if we (get it right), this could be the (way forward for a range of employees) on (different types of courses)...

Process Two: Look for Opposites

This text is about the considerations regarding whether / where to attend the course e.g. …(*attend*) ...(*do not attend*). One pole of the first concept is to *attend* the course. The opposite pole, which is *not* to attend, provides a meaningful comparison. The second pole of *attend* (i.e. *not attend*) is not found until near the end of the passage. If the second pole is not apparent, then in oral situations, it may be possible to clarify the problem by asking directly what the alternative course of action might be.

Mapping is of interest due to its ability to enhance understanding of a problem situation and hence the decision regarding whether and how to act or intervene in it. In order to encourage the analysis to be action orientated, a concept could usefully be expressed imperatively, highlighting actors and actions where possible. Verbs may sometimes be implicit where concepts are lengthy to express. Where possible, retain the original language of the problem owner and the names of any actors in the text, to retain ownership of the analysis and the solution. As an example the original phrases are now translated into concepts that can be interpreted into map form:

> (attend) ...(not attend)
> (in London) ...(in Plymouth) ...(split between Plymouth and London)
> (more convenient [for participants])
> (higher accommodation costs as we don't live there)
> (local staff)...
> (higher travelling costs for staff)
> (local expertise to tap)... (wider range of options)
> (much pleasanter place to be)
> (anybody who has studied at Plymouth before)
> (more intimate environment)... (outside of Plymouth)
> (fully satisfy all staff)

(incur the administrative burden of finding accommodation and paying fees)
(experience has shown)
(funding support for employees' attendance [at Plymouth])
[fail to keep up to date]... (get it right)
(business will suffer)
(the way forward for a range of employees)...
(these short courses)... (different types of courses)

Process Three: Graphically, Order and Link the Concepts

This process involves linking concepts together to show the line of reasoning adopted in the hierarchical process towards defining goals. For example, the decision *to attend (or not) a course in London*, results in *higher accommodation costs as we don't live there.*

Higher accommodation costs as we don't live there

attend (or not) a course in London → *Higher accommodation costs as we don't live there*

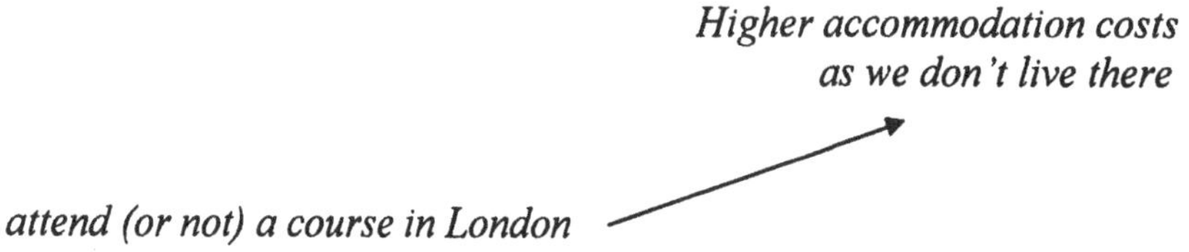

Because the *higher accommodation costs as we don't live there* depend on whether to *attend the course in London or not*, it is a super-ordinate outcome of the preceding concept, which is a subordinate option. An arrow pointing from one concept to another implies that it is one means towards achieving a desired end, or super-ordinate concept.

Where possible, use action-oriented statements to define concepts, but avoid normative statements such as 'should', 'need', 'ought to', or 'want'. *We need to attend a course in London* would be defining a desired option, not an action. The real action relates to finding ways to *fully satisfy the employees.*

Generic concepts, are those concepts that may be achieved through several different means, and are super-ordinate to specific items. The concept of *administrative burden of finding accommodation / paying fees* depends on whether they *attend in London or not* with the former in turn dependent on whether they have *funding support for employees' attendance [at Plymouth].*

administrative burden of finding accommodation / paying fees

↑

attend in London or not

↑

funding support for employees attendance [at Plymouth]

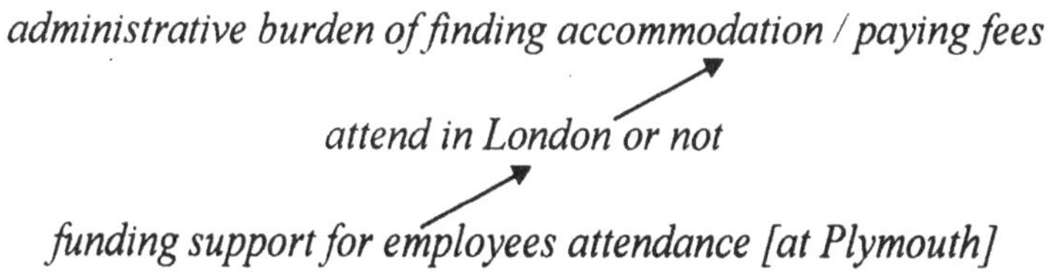

A negative link is implied between the concepts of *funding support for employees attendance at Plymouth* and *attend in London.* If our actions take into account that if we gain *funding support for employees attendance at Plymouth*, it implies a link with the second stated pole of the *attend in London* concept... we don't.

However, the first pole stated about *attend / not attend short courses*, is a bipolar concept, which is one where both poles are eventually explicitly stated in the text. Where people have a negative view of their world, they often use negative links in their cognitive maps, while those with much clearer goals, and a positive outlook, use more positive links. The problem owner's view of the world should be used to define the first pole of each concept (e.g. *business will suffer*).

Sometimes, concepts are not linked in with the rest of the map, and they form islands. This may be an important clue as to how the problem owner views the issue. In interviews, the researcher may explore the issue and its links further, but in written sources, some unstated links may need to be inferred. In this example, a concept of *keep up to date* was inferred as a link, between *attend short courses* and whether the *business will suffer.*

Typically, mapping should be done on one A4 sheet of paper (Figure 6.1), using pencils to map concepts, starting about one third of the way down the page. Text is best written in small blocks rather than across the page.

The Role of Computers in Cognitive Mapping

Practical applications of computer technologies in a mapping context have traditionally been more concerned with knowledge acquisition and structuring rather than knowledge representation, which is our present concern, although many of the methodological issues are pertinent to both contexts. In terms of concept mapping (Hammond, 1996) the use of nodes to denote concepts and links to show their relationships is familiar, as is the tension between providing users with a completely open initial structure, and one with varying degrees of skeletal structure or content.

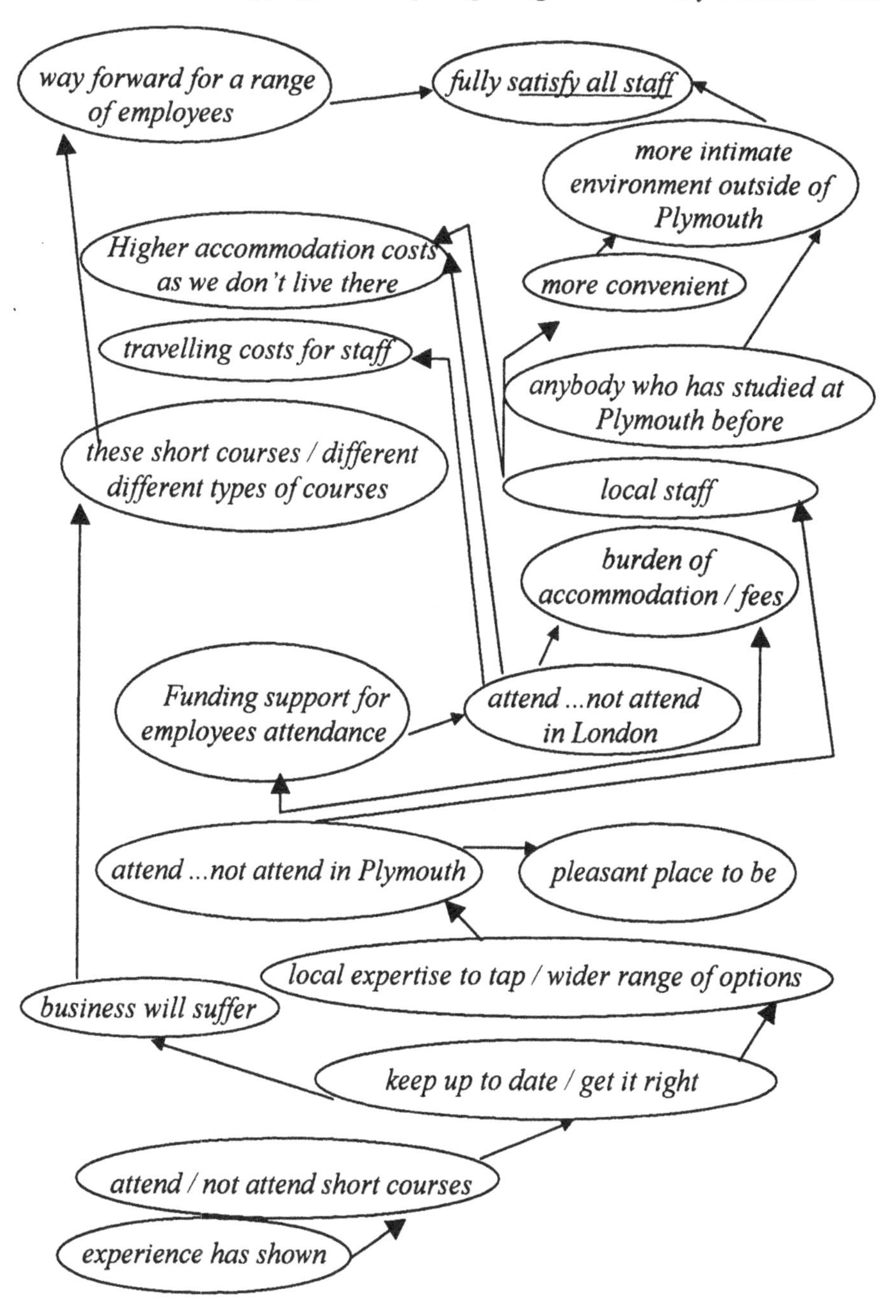

Figure 6.1 A cognitive map of the course attendance problem
Source: the author

Support for employing a more directed mapping environment is provided by Reader and Hammond's (1994) comparison of constrained and unconstrained knowledge mapping tools. Working in an unstructured environment, they found that the maps of students may become devoid of any structure, may focus on irrelevant aspects of the information, or may become idiosyncratic. In a more structured environment, maps were in closer accord with the goal of the learning task, and succeeded in communicating the main points of the information more effectively. Also, constrained tools resulted in more complete representations of the information than unconstrained tools.

Quantifying Differences between Cognitive Maps

Since these relatively early applications, it has been recognised that a family of techniques has developed, involving varying degrees of quantification of causal relationships between concepts. Although traditional cognitive mapping techniques tend to more qualitative approaches (Eden, 1990; Eden and Ackermann, 1993), systems dynamics modelling employs quantifiable functions, with explicit time considerations (Wolstenholme, 1994). Contingent on these developments, attempts have been made to incorporate a time dimension in measuring differences between cognitive maps (Langfield-Smith and Wirth, 1992), by a snapshot representation of maps as a directed network of an individual's beliefs in a particular arena. The quantitative comparison of both changes in individual maps over time, and the comparison between individuals at a particular point in time, makes this approach invaluable. Other attempts (e.g. Wang, 1996) to analyse the dynamic aspects of target systems represented as neural network models of cognitive maps are inappropriate as a means of representing knowledge, which is the main concern in this study.

Differences between cognitive maps may relate either to the content or to the structure of the maps that are being compared. Content difference is concerned with the variation in issues or elements contained within two or more cognitive maps, and variations in the causal beliefs between them. Structural differences relate to varying degrees of complexity of the map structure (Dunn et al, 1986). Given that cognitive maps represent the links between the causal beliefs of an individual pertaining to some issue, it is imperative that any attempt to quantify such links must still preserve the meaning inherent in the original belief system. Langfield-Smith and Wirth (1992) noted three main types of differences between cognitive maps:

1. Some aspects of an issue, defined by the 'elements' of it which are considered to be important by one individual, may not even be considered as being relevant by a different individual. It is conceivable that no common elements might be defined in a comparison between the adjacency matrices of two individuals.
2. Where two individuals both consider one element to be relevant to a particular issue, the set of beliefs that they both hold about the element may differ. An individual for whom a particular belief about the element is pertinent would record a non-zero score in a particular cell of the adjacency matrix, compared with a zero for the individual who discarded the particular belief.
3. Where merely the strength of two individual beliefs about an element differ, the corresponding cell in their adjacency matrices would record differing non-zero values.

Eight of the matrix distance measures proposed by Langfield-Smith and Wirth (1992) are defined in detail in Appendix 3. Formula 1 (F1) sums the absolute differences in scores between corresponding pairs of cells in two adjacency matrices. Initially, where only causal links between beliefs are considered, individual cells in the adjacency matrix took on values of +1, 0 or -1, defining a maximum distance of 2 in any cell of the distance matrix.

In cases where the cognitive maps of some individuals may include more elements in them than others, the matrix distance (Formula 2, F2) is weighted. Weights are based on the number of cells in each matrix that could be occupied, the unoccupied cells on the diagonal and the maximum difference between corresponding cells in the matrices.

A further refinement accounts for elements that either do not impact on, or on which no other elements have an impact. A transmitter was defined as an element which has an impact on at least one other element in a map, but on which, no other element has an impact. A receiver is an element which has no impact on any other element, but on which at least one other element has an impact. Formula 3 (F3) accounts for transmitters and receivers, and may be adjusted (Formula 4, F4) where there are a differing number of elements in each map.

When adjusted for the differing strengths of beliefs, raw elements of the adjacency matrix vary from -3 to +3, producing distance ratio Formula 5, (F5). When amended for cases where the two maps may contain different sets of elements Formula 6, (F6) accounts for elements which may be either common to the two maps, yielding a maximum value of + 6 in the distance matrix, or unique, with a maximum distance of + 1.

In Formula 7 (F7), the matrix distance formulae are broken down into those elements that are related to unique beliefs and those which are related to common beliefs. The later includes either those where the strength of commonly held beliefs varies or those where there may or may not be common beliefs with regard to common elements. Finally, this can be adjusted for varying strengths of the unique beliefs (Formula 8, F8).

Mapping the Decision to Study Shipping and Logistics

Based on transcripts of the focus groups which were set up to explore how a range of postgraduate students had perceived that their decisions to study International Shipping and International Logistics at Plymouth had been made, a questionnaire instrument was devised. This incorporated nine major issues (Figure 6.2) and several detailed items within each issue (see Chapters 4 and 5). Subsequent groups of students of international logistics and shipping were asked to complete these questionnaires, before being requested to map and identify the perceived links between these factors. These were then transcribed initially into a simple directional valency matrix (Figure 6.3), with cell values of +1 representing a positive link, 0 no link and -1 a negative link. Later, the strength of links was shown in an adjacency matrix, with cell values showing strong, moderate, weak or no links (3, 2, 1 or 0 respectively) in addition to the directional sign. This device of using a snapshot representation of maps as a directed network of an individual's beliefs enabled differences between the cognitive maps of individual students to be quantified by comparing pairs of these matrices.

Data was collected from several cohorts of students, involving combinations of shipping and logistics interests over a two-year period. One comparison of 38 students generated 703 map comparisons, another with 47 students generated 1081 comparisons, and one with 54 students created 1431 comparisons. Some of the measures proposed by Langfield-Smith and Wirth (1992) comparing the distances between maps were then computed for these groups, and analysed statistically although some proved redundant, given the mapping procedures employed. Formula F2, appropriate to comparisons between pairs of maps in which the number of elements in each map differed, was redundant where a standardised overall map structure was proposed, based on a fixed number of elements.

A 'mind map' linking each factor which was an issue in the decision to study at Plymouth using arrows to show the direction of cause and effect.

The strength of each link is numbered as:
1 = weak 2 = moderate 3 = strong, with a positive sign for positive links, and a negative sign for negative links.
Note: L/S = Logistics or Shipping

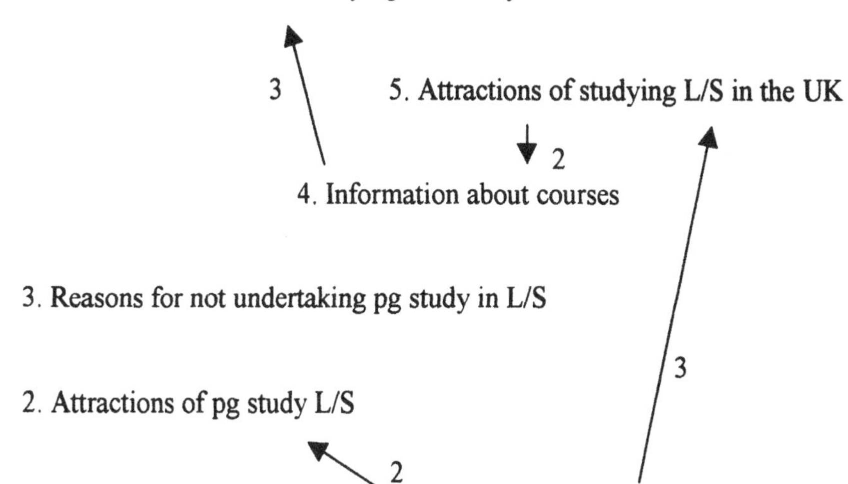

Figure 6.2 A hypothetical student's cognitive map, valency and adjacency matrix

Source: the author

A valency matrix showing the direction of each link is labelled as:
+1 is a positive link (i.e. A affects B positively);
-1 is a negative link (i.e. A affects B inversely).

Code	1	2	3	4	5	6	7	8	9
1. Family or friends		1			1				
2. Attractions pg study									
3. Not pg study									
4. Course information						1			
5. Attractions of study in UK				1					
6. Attractions of study in Pl							-1		
7. Why not study in Pl									
8. Why not study elsewhere									
9. Barriers to study at Pl									

An adjacency matrix showing how strong each link is labelled as:
1 = weak 2 = moderate 3 = strong.
+3 is a strong positive link (A affects B positively and strongly);
-2 is a moderate negative link (i.e. A affects B moderately and inversely).

Code	1	2	3	4	5	6	7	8	9
1. Family or friends		3			2				
2. Attractions of pg study									
3. Not pg study									
4. Course information						3			
5. Attractions of study in UK				2					
6. Attractions of study in Pl							-2		
7. Why not study in Pl									
8. Why not study elsewhere									
9. Barriers to study at Pl									

Figure 6.3 A hypothetical valency and adjacency matrix
Source: the author

Differences Between the Decision Maps of Individual Students

Langfield-Smith and Wirth's (1992) measures (Formulae F1 to F8) were computed for individuals and subsets of the Masters cohorts of 1997, 1998 and 1999. Analysis proceeded by assessing descriptive and inferential statistical results for each formula, and cohort. Groupings for which statistics were calculated included:

1. For each individual student, represented as the origin of a map pair, descriptive statistics of their overall individual pair-wise comparisons with the maps of all other students in their cohorts.
2. A single aggregate descriptive statistic for the whole group computed over all individuals in a cohort, based on the complete set of pair-wise comparisons between individuals in the cohort.
3. Descriptive statistics computed over all individuals in subgroups displaying particular characteristics, (e.g. common academic discipline, gender, age or nationality) represented as an originating map-set, comparing their map-set with the map-sets of all other members of their cohort.
4. Descriptive statistics computed over all individuals in subgroups displaying particular characteristics, (e.g. common academic discipline, gender, age or nationality) represented as an originating map-set, comparing their map-set with the map-sets of other subgroups in their cohort.

Inferential statistical tests included:

1. Tests of the normality of these statistical distributions.
2. Tests for significant differences between the measures of central tendency of these overall distributions.
3. Tests for significant differences between the measures of central tendency of subgroups within these distributions.

In subsequent tables, the degree of statistical confidence with which results are reported is represented as * (95 per cent), ** (99 per cent), *** (99.9 per cent) and **** (99.99 per cent).

Descriptive Statistics of the Distributions of Matrix Distance Measures

Table 6.1 summarises the analysis of the means of computed formulae values as calculated for each individual, as the originator of the map sequence. In the 1999 data, a mean of 1.33 represents the mean of 54 individual means for F3, in turn varying from an individual low of 0.62 to a high of 2.97. The individual means were based on the mean of formula scores between each individual as the originator in map comparisons and 53 other individuals. Table 6.2 gives further details by discipline.

Coefficient of variation data show that the 1997 data set was the least variable, followed by the 1999 data, but with the standard deviations of many formulae in the 1998 data set being so variable as to exceed their mean. The means of some formulae are relatively stable between years, namely the relatively simple measures F3, which ranged from 1.16 to 1.33 and F5 that ranged from 0.46 to 0.71. The compound or squared measures were more variable including F4 ranging from 0.70 to 3.72, F6 from 0.34 to 1.27, F7 from 0.62 to 2.64 and F8 from 0.48 to 2.27. It is also interesting that the simple distance measures failed to show the variability in the 1998 data set, which is much more apparent in the squared elements of other formulae. These values are indicative, and point to factors such as subject group, gender, age or nationality which are influencing the variation, rather than any cohort specific factors.

Are Computed Measures of Map Comparisons Normally Distributed?

In order to select appropriate statistical tests that might indicate any systematic sources in the variability of these formulae, assumptions about the normality of their distributions must be made, after testing the available evidence. The statistical distributions of this data for each cohort are reported in Table 6.3, showing the confidence with which a null hypothesis of no significant difference between the observed and normal distributions based on a Chi squared goodness of fit test could be rejected. The number of cases in which each individual student, represented as the originating map in pair-wise comparisons with all other students in the cohort, generated distributions with varying degrees of normality, are shown.

Table 6.1 Descriptive statistics of the means of each individual as a map originator for each comparative formula

Data shows the cohort sample size (n), mean, standard deviation (SD), minimum (min), maximum (max) and coefficient of variation (covar).

	Mean	SD	Min	Max	Covar
1997	n=38				
Formula 3	1.17	0.36	0.29	1.93	0.31
Formula 4	0.70	0.24	0.29	1.54	0.34
Formula 5	0.71	0.28	0.18	1.63	0.39
Formula 6	0.34	0.15	0.14	0.77	0.44
Formula 7	0.62	0.28	0.27	1.37	0.45
Formula 8	0.48	0.19	0.22	0.99	0.40
Distance	41.72	5.65	31.84	56.95	0.14
1998	n=47				
Formula 3	1.16	1.22	0.16	14.11	1.05
Formula 4	3.72	7.13	0.15	42.67	1.91
Formula 5	0.58	0.47	0.06	5.08	0.81
Formula 6	1.27	2.06	0.02	15.00	1.62
Formula 7	2.64	4.31	0.06	28.40	1.63
Formula 8	2.27	3.70	0.04	28.20	1.63
Distance	36.87	24.45	10.00	142.00	0.66
1999	n=54				
Formula 3	1.33	0.41	0.62	2.97	0.31
Formula 4	1.94	1.32	0.33	5.15	0.68
Formula 5	0.46	0.14	0.21	1.03	0.30
Formula 6	0.92	0.59	0.09	2.42	0.64
Formula 7	1.91	1.22	0.19	5.38	0.64
Formula 8	1.63	1.01	0.17	4.35	0.62
Distance	28.25	5.32	20.19	45.75	0.19

Source: the author

In none of the cohorts, were the aggregate distributions for any one measure taken over all individuals, normally distributed. In the 1997 data set, the aggregate simple distance measures (Formula 1) were however normally distributed for most individuals. Similar patterns of findings, even though frequently non-normal, were observed when comparing Formula 3 with Formula 4, and Formula 6 with Formulae 7 and 8.

Table 6.2 Descriptive statistics of variation in formulae by discipline

Data shows the minimum (Min), maximum (Max) and mean by discipline.

	Shipping			Logistics		
	Min	Max	Mean	Min	Max	Mean
F3: 1997	0.56	1.93	1.17	0.29	1.85	1.17
1998a	0.26	1.29	0.52	0.24	0.83	0.41
1998b	0.59	3.82	1.45	0.52	1.25	0.86
1999	0.78	2.97	1.34	0.62	2.16	1.33
F4: 1997	0.44	1.54	0.74	0.29	1.37	0.65
1998a	0.17	7.58	2.13	0.18	5.09	2.13
1998b	0.38	22.10	4.25	0.30	8.38	3.24
1999	0.36	4.40	1.74	0.33	5.15	2.14
F1: 1997	36.2	53.8	42.2	31.84	56.95	41.15
1998a	24.5	120.3	35.8	20.87	37.39	27.61
1998b	26.3	114.2	42.8	26.43	37.07	30.37
1999	20.3	45.8	27.5	20.19	38.13	29.15
F5:1997	0.23	1.75	0.71	0.18	1.63	0.72
1998a	0.20	1.75	0.56	0.24	0.90	0.47
1998b	0.27	1.74	0.67	0.28	0.69	0.48
1999	0.26	1.03	0.47	0.21	0.72	0.45
F6:1997	0.14	0.77	0.36	0.15	0.57	0.31
1998a	0.10	6.75	1.55	0.13	4.09	1.66
1998b	0.12	4.31	1.37	0.11	2.84	1.20
1999	0.13	2.13	0.81	0.09	2.42	1.06
F7:1997	0.27	1.37	0.67	0.27	1.02	0.56
1998a	0.23	16.57	3.38	0.27	9.25	3.54
1998b	0.25	10.27	2.93	0.21	5.44	2.41
1999	0.25	5.19	1.67	0.19	5.38	2.19
F8:1997	0.23	0.99	0.51	0.22	0.77	0.43
1998a	0.19	10.72	2.50	0.24	6.61	2.69
1998b	0.20	8.52	2.53	0.19	3.96	2.04
1999	0.21	4.35	1.46	0.17	3.76	1.82

Source: the author

This suggests some association between the various measures, as might be expected given their common derivations. However, many distributions were non-normal, even for individuals. In the 1998 data Formula 5 generated the best approximations to normality but most measures were mainly non-normal. In the 1999 cohort measures 1 and 3, and to a lesser extent, 4 and 5 displayed a better fit to normality than in the 1998 cohort, but particularly the compound measures, F6, F7 and F8 were non-normal. Further analysis of the sources of variation was the next priority.

Table 6.3 Distributions of pair-wise map comparison formulae

Observed distributions of pair-wise map comparison formulae. Number of cases, taking each individual as an origin, generating normal distributions, or non-normality at the level of confidence shown (e.g. 95%) based on a Chi squared goodness of fit test.

	F3	F4	F1	F5	F6	F7	F8
1997, n = 37							
Normal	21	25	34	22	21	22	21
95	6	4	1	6	5	5	3
99	5	2	0	6	5	5	7
99.9	2	1	0	2	3	1	2
99.99	3	5	2	1	3	4	4
1998, n = 47							
Normal	11	12	3	29	17	14	18
95	11	3	3	6	3	8	10
99	6	5	12	7	6	6	1
99.9	6	2	10	2	6	3	6
99.99	13	25	19	3	15	16	12
1999, n = 54							
Normal	45	33	51	40	17	15	18
95	6	8	2	6	7	8	4
99	3	8	1	4	6	10	13
99.9	0	0	0	3	7	3	5
99.99	0	5	0	1	17	18	14

Source: the author

In attempting to explain variations in comparisons between maps, the mean of the set of values for a particular formula for one originating individual was computed in relation to all other individuals. Originating individuals were then grouped together on the basis of some commonality, including academic discipline, gender, age and origin, for each cohort of students. Aggregate values of each formula for the various subgroups were then computed. Non-parametric measures of differences in the central tendency of the distributions of the various formulae were then attempted, for these subgroups. Table 6.4 shows the results of these statistical tests.

Table 6.4 Significant differences in the central tendency of distributions of map comparisons between each subgroup and the whole cohort

Significant differences based on two-sample Mann-Witney U tests, between the central tendency of distributions for each subgroup and the whole cohort, based on the means of map comparisons between each individual as an originating map, and all others.

	Subject	Gender	Age	Origin
Formula 3	No significant differences in the 1997 data set			
1998	*		****	****
1999			**	
Formula 4	No significant differences in the 1997 data set			
1998				*
1999				*
Formula 1	No significant differences in 1997 or 1999 data sets			
1998			****	****
Formula 5	No significant differences in the 1997 data set			
1998				****
1999			*	
Formula 6	No significant differences in 1997 or 1999 data sets			
1998			***	
Formula 7	No significant differences in 1997 or 1999 data sets			
1998			***	
Formula 8	No significant differences in 1997 or 1999 data sets			
1998			****	

Source: the author

Explanations of Differences in Formulae Means, between Individuals

The analysis of raw means for originating maps showed relatively few significant differences between the two groupings used within each category. No variation at all was observed by gender and the only case of significant variation was noted by subject discipline in the 1998 cohort. On the age and nationality factors, the 1998 data set again showed more variability than the others, with no significant differences being observed in the 1997 data set, and only three instances in the 1999 data set. The formulae most subject to variation were F3 and F1, the relatively raw comparisons. Here, the non-European students, possibly older and with more work experience or differing familiarities with the English language, displayed different views to the other groups.

However, this approach did not address the issue of map comparisons between similar pairs of individuals at both the origin and destination ends of the comparison, and it was necessary to re-examine these approaches. Groups of measures were recomputed on the basis of types of paired map comparisons, and the results are shown below.

Academic Discipline, Gender, Age, Nationality and Map Comparisons

In this analysis, the pairings of both the origin and destination maps were controlled, enabling subgroups of for example gender to be identified as comprising male to male, male to female and female to female comparisons. Analyses of the impacts of academic discipline, gender, age and nationality differences on differences between pairs of map comparisons are presented in Tables 6.5 to 6.8. The statistical significance of differences between the measures of central tendency of distributions of Langfield-Smith and Wirth's (1992) quantitative comparisons between maps were estimated using non-parametric tests, given the non-normality of many distributions. As with comparisons of the simple means, the significance of non-parametric Mann-Witney U test (U) results for differences in the measures of central tendency between pairs of subgroups within the overall samples are shown, along with the Kruskal Wallis (KW) tests for multiple comparisons between subgroups.

Table 6.5 Differences between the measures of central tendency of distributions of formulae by discipline

Paired views, matched between subgroups showing the significance of non-parametric two-sample (e.g. S/B) and multiple sample comparisons (Overall) : * = 95% ** = 99% *** = 99.9% **** = 99.99%.

1997: Discipline
Sample size: (S)hipping 210, (B)oth 357, (L)ogistics 136

F3	
F4	S/B *, S/L **, Overall *
F1	
F5	
F6	S/L **, Overall *
F7	S/L **, Overall *
F8	S/L **, B/L *, Overall *

1998: Discipline
Sample size: (S)hipping 253, (B)oth 552, (L)ogistics 276

F3	S/B ****, S/L ****, B/L ****, Overall ****
F4	S/B ***, S/L****, B/L **, Overall ****
F1	S/B ****, S/L ****, B/L ****, Overall ****
F5	S/B *, S/L ****, B/L ****, Overall ****
F6	S/L *, B/L ***, Overall **
F7	S/L ***, B/L ****, Overall ****
F8	S/L *, B/L ***, Overall ***

1999: Discipline
Sample size: (S)hipping 406, (B)oth 725, (L)ogistics 300

F3	
F4	S/B **, S/L ****, B/L *, Overall ****
F1	S/B ****, S/L ****, B/L **, Overall ****
F5	S/L **, Overall *
F6	S/B ***, S/L ****, B/L ***, Overall ****
F7	S/B ***, S/L ****, B/L ***, Overall ****
F8	S/B ***, S/L ****, B/L ***, Overall ****

Source: the author

Table 6.6 Differences between the measures of central tendency of distributions of formulae by gender

Paired views, matched between subgroups showing the significance of non-parametric two-sample tests (e.g. M/B) and multiple sample comparisons (Overall): * = 95% ** = 99% *** = 99.9% **** = 99.99%.

1997: Gender
Sample size: (M)ale 377, (B)oth 281, (F)emale 45

F3	M/B ****, M/F **, Overall ****
F4	M/B ****, M/F ***, Overall ****
F1	
F5	M/B ***, M/F **, Overall ***
F6	M/B ****, M/F ****, Overall ****
F7	M/B ****, M/F ****, Overall ****
F8	M/B ****, M/F ****, Overall ****

1998: Gender
Sample size: (M)ale 903, (B)oth 172, (F)emale 6

F3	M/B ****, Overall ****
F4	M/B ****, M/F ***, Overall ****
F1	M/B ****, Overall ***
F5	
F6	M/B ****, M/F ***, Overall ****
F7	M/B ****, M/F ***, Overall ****
F8	M/B ****, M/F **, Overall ****

1999: Gender
Sample size: (M)ale 946, (B)oth 440, (F)emale 45

F3	M/B **, M/F **, Overall ***
F4	M/B ****, M/F ****, B/F *, Overall ****
F1	M/B ***, M/F **, Overall ***
F5	M/B **, Overall **
F6	M/B ****, M/F ****, Overall ****
F7	M/B ****, M/F ****, Overall ****
F8	M/B ****, M/F ***, Overall ****

Source: the author

Table 6.7 Differences between the measures of central tendency of distributions of formulae by age

Paired views, matched between subgroups showing the significance of non-parametric two-sample tests (e.g. Y/B) and multiple sample comparisons (e.g. Overall): * = 95% ** = 99% *** = 99.9% **** = 99.99%.

1997: Age
Sample size: (Y)ounger (< 30 years) 528, (B)oth 165, (O)lder 10

F3	
F4	Y/B *, Overall *
F1	
F5	
F6	Y/B ****, Y/O *, Overall ****
F7	Y/B ***, Y/O **, Overall ****
F8	Y/B ****, Y/O *, Overall ****

1998: Age
Sample size: (Y)ounger 666, (B)oth 370, (O)lder 45

F3	Y/B ****, Y/O *, Overall ****
F4	
F1	Y/B ****, Y/O **, Overall ****
F5	
F6	Y/B *, Y/O **, Overall ***
F7	Y/O *, Overall ****
F8	Y/B *, Y/O **, Overall **

1999: Age
Sample size: (Y)ounger 1035, (B)oth 368, (O)lder 28

F3	
F4	
F1	Y/B **, Overall *
F5	Y/B **, Overall *
F6	
F7	
F8	

Source: the author

Table 6.8 Differences between the measures of central tendency of distributions of formulae by nationality

Paired views, matched between subgroups showing the significance of non-parametric two-sample tests (e.g. E/B) and tests for multiple sample comparisons (Overall): * = 95% ** = 99% *** = 99.9% **** = 99.99%.

1997: Nationality

Sample size: (E)uropean 528, (B)oth 165, (W)orldwide 10

F3	
F4	
F1	E/B ***, E/W *, Overall ***
F5	
F6	
F7	Overall *
F8	

1998: Nationality

Sample size: (E)uropean 465, (B)oth 496, (W)orldwide 120

F3	E/B ****, E/W ****, B/W ****, Overall ****
F4	E/B *, E/W **, Overall **
F1	E/B ****, E/W ****, B/W **, Overall ****
F5	E/B ****, E/W ****, Overall ****
F6	
F7	E/B *, Overall *
F8	E/B *

1999: Nationality

Sample size: (E)uropean 1035, (B)oth 368, (W)orldwide 28

F3	
F4	E/B ****, E/W ****, B/W **, Overall ****
F1	
F5	
F6	E/B ****, E/W ****, B/W *, Overall ****
F7	E/B ****, E/W ****, B/W **, Overall ****
F8	E/B ****, E/W ****, B/W *, Overall ****

Source: the author

Summaries of the statistical variations between subgroups of cohorts in Langfield-Smith and Wirth's (1992) formulae were computed based on subject, gender, age and origin, for the 1997, 1998 and 1999 cohorts. Table 6.9 summarises the instances where differences between measures of the central tendency of the distributions of sub-sets of samples, within any one factor and student cohort, were significant at least at the 99.9 per cent level based on Kruskal Wallis (KW) tests of differences.

Row subtotals by formula, and column subtotals by factor reveal some of the main sources of variation in the measures. Differing academic disciplines and gender presented larger sources of variation than changes in nationality, with age generating the least variation of all. Gender differences presented a consistently significant source of variation between pairs of maps, in 18 out of 21 possible cases. In terms of variations by formulae, there was relatively little difference, except that Formula 5 displayed fewer differences between subgroups than the other measures. Overall totals for each cohort, including 10 for 1997, 18 for 1998 and 15 for 1999 again showed the greatest variability in the 1998 data and the least in 1997.

In some cases the Mann-Witney U test results between sample subgroups should be treated with caution, due to small sample sizes. The following instances are noteworthy:

1. Regarding gender, female to female comparisons were 45 in 1997, 6 in 1998 and 45 in 1999.
2. In the age subgroup, mature to mature was only 10 in 1997 and 45 in 1998, with young to young only 28 in 1999.
3. In the nationality subgroup, there were only 10 worldwide to worldwide comparisons in 1997 and 28 in 1999.

Taking each subgroup in turn, in an attempt to explain the main sources of variation yielded the following effects:

1. By subject, consistently over all three cohorts, formulae F4, F7 and F8 provided the best discriminators. Single subgroup combinations of shipping and logistics showed the greatest contrast in 1997 and 1999, but with the 'both' and logistics combination producing the most contrast in 1998.
2. By gender, over all three cohorts, formulae F4, F6, F7 and F8 provided consistently good discrimination, with consistently good contrast on the male and 'both' subgroup combinations.

Table 6.9 Multiple comparisons between subgroups of measures of the central tendency of distributions of formulae by cohort

Differences significant at least at the 99.9% level are shown for Kruskal Wallis (KW) tests of differences between sample sub-sets.

	Subject	Gender	Age	Origin	Total
Formula 3					6 / 12
1997		*			
1998	*	*	*	*	
1999		*			
Formula 4					6 / 12
1997		*			
1998	*	*			
1999	*	*		*	
Formula 1					7 / 12
1997				*	
1998	*	*	*	*	
1999	*	*			
Formula 5					3 / 12
1997		*			
1998	*			*	
1999					
Formula 6					7 / 12
1997		*	*		
1998		*	*		
1999	*	*		*	
Formula 7					7 / 12
1997		*	*		
1998	*	*			
1999	*	*		*	
Formula 8					7 / 12
1997		*	*		
1998	*	*			
1999	*	*		*	
Total	11 / 21	18 / 21	6 / 21	8 / 21	43 / 84

Source: the author

3. By age, no single formula, and no single subgroup comparison yielded consistently good measures of the contrasts between formulae or subgroups over all three cohorts.
4. Finally, by origin, no consistency between subgroups was apparent. Only formula F1 revealed any variation in 1997, and whilst F3, F1 and F5 provided variation between subgroups in 1998, they provided no variation in 1999.

Taken overall, these results indicate that:

1. Raw map distance measures based on particular individuals as origins were of limited value in attempting to explain sources of variation between them.
2. Gender and possibly academic discipline appeared to present the most likely sources of variation between measures.
3. Substantial redundancy was present if all measures were analysed in a particular context. In particular the compound formulae, measures F6-F8 were often related, as were the simple measures F3 and F4.
4. Attempts to cluster students into particular categories are unlikely to be fruitful, as no systematic or stable, grouping criterion was readily apparent.

Student Reactions to the Mapping Approach

Following the initial survey the cognitive mapping exercise was repeated with a review questionnaire being administered to all students in one cohort. Some ethical concerns remain regarding the principle of attempting to reduce a potentially complex cognitive map into a matrix format. Does this constrain the variety of elements which individuals might include in their maps? Does an imposed spatial contiguity mapping of issues encourage false relations to be inferred between them? Does reduced space for inclusion of additional comments, and imposition of words for key concepts deny their construct validity? Although the experiment was justified on the basis of the high construct validity of the initial focus groups, respondents in the review survey were also asked how they felt about reducing their maps to a matrix in this way. Table 6.10 shows that of 43 responses received, 15 were positive, 20 were neutral and 8 were negative, with only three of the respondents feeling a need to express this concern directly. Overall the benefits outweighed the costs, in terms of the initial acceptability of the approach.

Table 6.10 How do you feel about reducing a cognitive map to matrix format?

Response	Frequency
Positive, which included:	15
It clarifies understanding	4
It is interesting	2
It is a good concept	2
A logical way of presenting data	2
A quick and economical way of presenting data	2
It gives a reduced overview	1
It is very useful	1
It is organised	1
Neutral, which included:	20
Passive / OK	14
I am not yet sure	2
Its a new experience	1
It just had to be done	1
Its fine if there are not too many connections	1
What a bizarre question	1
Negative, which included:	8
It takes away the full effect of the mapping	3
Its confusing	2
Drawing the map is difficult	2
It is time consuming	1

Source: the author

Additional educational benefits of this approach have been noted elsewhere (Dinwoodie, 1999c). At the lowest level, merely asking students to reflect on the reasons why they embarked on their current course of study can stimulate an important process of vocational reflection (Kidd and Killeen, 1992). At a higher level, by associating an important personal issue with other parts of the curriculum, it can raise the motivation to study associated disciplines, with the practical use of cognitive mapping presenting an obvious case in point.

Conclusion

In this chapter, a technique was presented which enabled the structure of the decisions made by individual vocational students to embark on postgraduate courses in international shipping and logistics, to be analysed in more detail. The technique of cognitive mapping also enabled students to express how they felt that a range of issues in their individual decisions had been inter-linked thus enabling them to outline their perceptions of the process whereby they had come to enrol on their courses. Causal relationships between key concepts in the decision process were quantified for each individual student, by representing their maps as a directed network of an individual's beliefs. Next, differences between the representations of the cognitive maps of individuals were estimated by using a range of descriptive formulae, which were then translated into computer programmes. Following this the measures were computed and comparisons made between individual maps. Data was collected for classes of postgraduate shipping and logistics students, over a three-year period, generating up to 1430 pair-wise comparisons between the individuals in a single cohort. Most students were unconcerned about being asked to reduce their maps to matrix format, despite some loss in information. In the process of computing Langfield-Smith and Wirth's (1992) measures, which compared the distance between maps expressed as valency matrices, the strength of comparisons represented in adjacency matrices were also computed.

In the first reported analysis of this type of data using this approach, much exploratory work, involving calculating and reporting descriptive statistics of the map comparison formulae, and the statistical distributions of the comparisons was undertaken. Although rather tedious prima facie, these initial computed comparisons revealed genuinely new insights into the structure of the decision process. Based on the work presented here, it is suggested that:

1. Distance measures constrained by the numbers of transmitters or receivers, and the strength of relationships where appropriate, probably yield the best discriminators.

2. The standardised mapping process adopted here generated some redundancy between map comparison measures. Typically, computation of F1, F4 and one of the measures F6-F8 would describe the major sources of variation between maps.
3. The degree of variation between individuals was not constant from year to year, necessitating samples drawn over several years.
4. Analysis of the raw means of map differences based on single individuals as the originating map generated less systematic variation than comparisons between similar subgroups of individuals.
5. Variation between subgroups on the basis of gender (male / female) and subject discipline (shipping / logistics) both accounted for significant differences in maps.

Overall, the approach generated a wealth of data that affords detailed comparison of maps between individuals and groups, both between individuals at the same time, and between groups over time. On the basis of these findings, further comparison and analysis of individual views, presented in the next chapter, will concentrate on differences in subject discipline as a source of variation.

7 Analysis and Explanation of Observations

Introduction

This chapter aims to discuss explanations of the observed patterns of views and maps presented thus far. Given the paucity of reported prior applications in this field of some of the approaches adopted, the analysis proceeds using a cumulative empirical approach. As a starting point, further descriptive comparisons are made between the various groupings within the cohorts that were surveyed. The discussion reports the factors that attracted particular subgroups of students to undertake study, noting some of the inter-correlations between them and the problems of attempting to build explanatory models. Further comparisons of cognitive maps are then made, including an attempt to explain the distance measures, noting the inter-correlations between comparative formulae scores. No single simple model emerged, either when attempting to build aggregate explanatory models of formulae scores, or when considering issue-based explanations of the pair-wise distance measures. In the light of this, an attempt was made to fit a known theoretical structure, with limited success. The chapter concludes by querying the wisdom of attempting to make broader theoretical statements within the current research context.

Analysis of Issues by Subgroups

How did the issues of concern vary between subgroups of shipping and logistics students when making their decisions to undertake study? Further applications of the bespoke instrument that was developed and analysed initially (see Chapter 5) provided some answers to this question. By way of convenience, Table 7.1 shows the summary codes of a range of descriptive variables used to define the main issues in this analysis. Their mean scores and rankings by subgroups, based on gender and academic discipline, are shown in Table 7.2 with subgroup details in Appendices 4.1 to 4.12. The data set based on 1997, 1998 and 1999 cohorts combined, showed the key issues for all groups to be the 'reasons for undertaking postgraduate study', the 'reasons for studying in the UK', and 'sources of information'. Of lesser importance overall were 'money', 'reasons for not studying at postgraduate level', 'barriers' and 'reasons for not studying elsewhere'.

Table 7.1 A summary of the issues that influenced the decision to study

Data shows issues, their statement number and code. How important were:
5: family or friends in influencing your decision? [Friends]
11: reasons that made the study of shipping at postgraduate level attractive? [Postgrad]
19: reasons that might have put you off studying at postgraduate level? [NotStudy]
27: sources of information about courses? [Info]
36: reasons that made the study of shipping in the UK attractive? [UK]
45: reasons that made the study of shipping at Plymouth attractive? [Plymouth]
65: reasons that might have put you off the study of shipping at Plymouth? [NotPlym]
75: reasons for not studying elsewhere? [NotElse]
82: barriers that might have put you off studying shipping at Plymouth? [Barrier]
83: money factors? [Money]
88: the quality of life in the city, your accommodation? etc. [City]
89: teaching methods? [Teaching]

Source: the author

The role of friends and the attractions of study in Plymouth were slightly higher for females, with barriers, money and teaching methods presenting less important issues. The importance of teaching methods was rated more highly by shipping students than logistics students.

Table 7.2 Mean scores and ranks of issues by subgroup using combined 1997-1999 data

Group	Males		Females		Logistics		Shipping	
Sample size	115		24		65		74	
Issue	Rank	mean	Rank	mean	Rank	mean	Rank	mean
Postgrad	1	1.83	1	1.71	1	1.78	1	1.84
UK	2	1.68	2	1.63	2	1.62	2	1.72
Info	3	1.55	3=	1.25	3	1.42	3	1.57
NotPlym	4	1.41	6=	1.17	4	1.26	5	1.46
Teaching	5	1.33	9=	1.08	10	1.05	4	1.50
Friends	6	1.32	3=	1.25	5	1.25	7=	1.36
Plymouth	7	1.31	5	1.21	7	1.20	6	1.38
NotElse	8	1.27	9=	1.08	6	1.22	9=	1.26
Barrier	9	1.23	6=	1.17	8	1.18	9=	1.26
NotStudy	10	1.20	11	0.96	11	0.92	7=	1.36
Money	11	1.03	8	1.13	9	1.11	11	1.00

Source: the author

Analysis of Popular Items by Subgroup

After considering variation in the broad issues, 20 of the more popular items within these issues were also analysed (Table 7.3) in an attempt to reveal any systematic variation in their overall importance in the decision to study. Variation in these items by gender and academic discipline is shown in Table 7.4. It is apparent that:

1. Career plans, a desire for new knowledge, and a recognised course affording good opportunities were the main attractions for all groups.
2. The importance of studying shipping was high to shipping students, but not to logisticians.
3. The desire to return for more study, important to Diploma students (mean score 1.25), was a less important concern for Masters students, but it is included as an important concern of experienced students.
4. The stated desire of the logisticians to go abroad, where many attended in the Netherlands and the UK, was greater than for shipping students.

Table 7.3 Items and their abbreviations as used in subsequent analysis

8: Talking to my family and friends influenced my decision to study. [Talk family]

Study of logistics or shipping at postgraduate level was attractive because I wanted:
12: to broaden my knowledge / learn new things. [New knowledge]
13: to enact my long term career plans. [Career plans]
15: to broaden my opportunities / be sure to find a job. [Opportunities]
16: to specialise in shipping, as I want to work there. [Shipping]
17: more study, following my work experience. [More study]
34: I found out about courses by talking to students on the course. [Talk students]

Study of logistics or shipping in the UK was attractive because:
37: to go abroad makes life more interesting. [Go abroad]
38: a British qualification is recognised worldwide. [Recognised]
40: I could practise English, the language of shipping, to gain more opportunities. [Practise English]
42: the UK academic system is more relevant to industry. [Relevant]

Study of logistics / shipping at Plymouth was attractive because:
50: the course specialises in shipping. [Specialist]
57: the university accepted me. [Accepted me]

I might have been put off study at Plymouth if I had found that the university:
66: gave me too little information. [Information]
68: was rated below other universities on its course quality, library etc. [Ratings]
69: had not offered me a place at Masters level. [Masters place]
73: students had given me bad reports about it. [Student Reports]

I might have put me off study of shipping or logistics at Plymouth if:
90: my first contacts with staff had not made me feel welcome. [Welcome]
91: if my expectation to learn and not merely to analyse statistics was not met. [To learn]
93: my expectation of a practical course with visits to companies and professional links was not met. [Practical course].

Source: the author

Table 7.4 Variation in the rank and mean scores of popular items by gender and discipline

Item	Male Rank / mean		Female Rank / mean		Shipping Rank / mean		Logistics Rank / mean	
Talk family	16	1.31	9	1.42	20	1.38	9	1.28
New knowledge	2	2.11	2	2.29	2	2.09	1	2.20
Career plans	1	2.14	3	2.25	1	2.22	2	2.09
Opportunities	4	1.92	5	1.92	7	1.85	5	1.94
Shipping	12	1.37	20	1.17	5	1.96	42	0.62
More study	40	0.91	31	1.00	31	1.15	41	0.68
Talk students	21	1.15	27	1.04	29	1.15	18	1.06
Go abroad	18	1.24	11	1.38	38	1.01	7	1.55
Recognised	3	1.90	1	2.33	6	1.95	3	2.00
Practise English	6	1.71	4	1.96	13	1.57	4	1.97
Relevant	19	1.23	26	1.04	17	1.43	28	0.92
Specialist	14	1.33	28	1.04	4	1.96	48	0.51
Accepted me	15	1.32	16	1.25	16	1.45	22	1.02
Information	11	1.37	7	1.50	14	1.55	12	1.20
Ratings	9	1.43	10	1.42	8	1.69	15	1.12
Masters place	5	1.82	6	1.83	3	1.97	6	1.64
Students reports	13	1.36	25	1.04	15	1.51	19	1.06
Welcome	8	1.43	12	1.33	12	1.58	10	1.23
To learn	7	1.44	8	1.42	9	1.66	13	1.18
Practical course	10	1.37	17	1.21	11	1.62	20	1.03

Source: the author

5. The desire to practise English was of greater importance to logisticians, many of whom were of Dutch origin, studying in a second language, rather than shipping students. The importance of a 'relevant' course was greater to shipping students.
6. The importance of a specialist and practical course in shipping was very high amongst shipping students and males, but much less so for logisticians.

Logically, almost all items that most strongly influenced the decision to study can be grouped into broader sets of motives, related to:

1. A desire to work in 'shipping' ('opportunities', 'shipping', 'practise English', 'specialist').
2. A desire to 'study' as an end in itself ('new knowledge', 'more study', 'specialist', 'to learn', 'practical course').
3. A desire to study as part of a 'personal' development process ('career plans', 'go abroad', 'accepted me', 'Masters place').
4. The influence of 'place' in the decision ('recognised', 'practise English', 'relevant', 'ratings').
5. The role of 'contacts' in the decision ('talk family', 'talk students', 'information', 'student reports', 'welcome').

Each of these broader sets, in turn, was mediated by length of prior relevant experience, gender and academic discipline.

In order to test the strength and nature of these relationships, correlation and regression analysis were employed. Spearman's rho non-parametric correlation coefficients, not dependant on assumptions of normality in data sets, revealed a range of inter-correlations (Table 7.5) between items in the broader sets, discussed below. However, parametric regression procedures were used to broadly calibrate possible relationships between items. Although these procedures are relatively robust in coping with non-normality in data sets, summarising the direction of relationships accurately, coefficients should be viewed as being merely indicative. Using a stepwise procedure, multiple regression of each of these dependent variables revealed relevant experience, discipline and gender, to be significant explanatory variables but gender was never so (Table 7.6).

A Desire to Work in the Shipping Industry

A desire to work in shipping was defined by a desire to broaden 'opportunities' through high-level study of it, an explicit desire to work in 'shipping', to 'practise English' as the language of shipping, and to chose a 'specialist' course in it. Statistically significant inter-correlations (Table 7.5) were found between the items:

1. 'Opportunities' and the other three items in this group.
2. 'Shipping' and 'opportunities' and 'specialist'.
3. 'Practise English' and 'opportunities'.
4. 'Specialist' and 'opportunities' and 'shipping'.

8	.25	.39	.21	.23	.16	.04	.07	.18	.10	.23	.27	.32	.34	.31	.36	.26			
5	.34	.22	.16	.25	.15	.09	.09	.05	-.01	.24	.17	.26	.40	.47	.33	.31	.53		
6	.12	.22	.22	.27	.17	.06	.07	-.05	-.05	.23	.26	.30	.45	.37	.38	.30	.44	.57	

Table 7.6 Explanations of key items in relation to experience, gender and discipline

Topic / Item	Explanatory variables and regression coefficients			
	Constant	Years of experience	Discipline	R^2
Interest in shipping				
Shipping	1.985		-1.370	.32
t value	16.411		-7.916	
Practise English	1.960	-0.071		.07
t value	16.646	-3.140		
Interest in study				
Specialist	1.985		-1.478	.40
t value	17.928		-9.328	
New knowledge	No variables entered			
Personal change				
Career plans	No variables entered			
Go abroad	1.275	-0.058	0.396	.11
t value	8.527	-2.710	2.195	
Accepted me	1.676		-0.753	.12
t value	13.397		-4.209	
Place of study				
Practise English	(see above)			
Ratings	1.735		-0.612	.08
t value	13.929		-3.436	
Contacts				
Talk family	No variables entered			
Talk students	No variables entered			
Welcome	1.573		-0.343	.03
t value	13.436		-2.046	

Source: the author

In view of these relationships, the items 'shipping' and 'practise English' were felt to succinctly, and mutually exclusively, cover most of the variation within this set of items.

As might be expected, Table 7.6 shows that the academic 'discipline' which a student was studying explained about one third of the variation in the importance of their interest in shipping and also offered a limited explanation of the desire to 'practise English'. Given that English is the international language of shipping, this was also as expected.

A Desire to Study as an End in Itself

Student desires to acquire new 'knowledge', undertake 'more study', find a 'specialist' course in shipping, 'to learn' and to make industrial contacts on a 'practical course' were felt to be indicative of a desire to undertake study as an end in itself. Inter-correlations between these items (Table 7.5) revealed that:

1. 'New knowledge' was only correlated with a desire 'to learn'.
2. 'More study' was correlated with 'specialist' and 'practical course'.
3. 'Specialist' was correlated with all items excepting 'new knowledge'.
4. 'To learn' was correlated with all other variables excepting 'more study'.
5. 'Practical course' was correlated with all the other items excepting 'new knowledge'.

Given that 'specialist' was related to all other variables except 'new knowledge', to which it was largely unrelated, these two items were selected for analysis.

The major explanation (about 40 per cent) of the desire to study as an end in itself was again provided by variation in the academic 'discipline' for which a student was registered. The distinction between being registered on a shipping or logistics course was the only statistically significant explanation of why students were attracted to a university offering a 'specialist' course in shipping.

A Desire to Study as Part of a Personal Development Process

The items relating to a desire to enact long term 'career plans', 'go abroad' for reasons of interest, and finding a university attractive because it 'accepted me', or offered a 'Masters place' all related to a desire to study as part of a process of personal development. The only inter-correlated items within this subset included:

1. 'Career plans' related to 'Masters place'.
2. 'Go abroad' unrelated to any other item, but almost to 'Masters place'.
3. 'Accepted me' related to 'Masters place'.

4. 'Masters place' related to 'careers plans' and 'accepted me' and almost to 'go abroad'.

'Masters place' was the only item excluded from further analysis.

Only the 'go abroad' and 'accepted me' items could be explained in statistically significant terms by the explanatory variables (Table 7.6). Subject 'discipline' alone 'explained' 12 per cent of the variation in the item 'accepted me', and even when combined with 'years of experience' accounted for only 11 per cent of the variation in a desire to 'go abroad'. The weak negative link between 'years of experience' and a desire to 'go abroad' reflects the weakening lure of adventure to more experienced students, and an increasing appeal of a more settled life that their new qualification might offer them.

The Influence of Place in the Decision

Geographical considerations influencing the decision concerning where to study were apparent where UK qualifications were perceived as being 'recognised' and the UK was also considered to be a good place to 'practise English' with industrially 'relevant' courses. More locally, the comparative institutional 'ratings' were also important. Inter-correlations between these items revealed that:

1. 'Recognised' was related to 'practise English' and 'relevant'.
2. 'Practise English' was correlated with 'recognised'.
3. 'Relevant' was related to 'recognised' and 'ratings'.
4. 'Ratings' was related to 'relevant'.

The items 'recognised' and 'relevant' were accordingly excluded from further analysis.

The desire to 'practise English' was weakly negatively related to the 'years of experience' (Table 7.5), indicating a greater desire to study in the UK among less experienced students, who may not have already had such opportunities. The importance attached to the ratings of particular institutions was also negatively related to the subject code used (shipping = 1, logistics = 2), reflecting a greater concern amongst shipping students for choosing institutions with highly rated courses or facilities.

The Role of Contacts in the Decision

Contact with other people was apparent in 'talking to family' and friends, 'talking to students' on the course, 'information' from the university, 'student reports' about a university and the 'welcome' apparent to applicants in initial contacts with staff. Inter-correlations between these items revealed that the importance of:

1. 'Talk family' was correlated with 'student reports' and 'welcome'.
2. 'Talk students' was related to 'student reports'.
3. 'Information' was related to 'student reports' and 'welcome'.
4. 'Student reports' were related to all other items.
5. 'Welcome' was related to 'information', 'student reports' and 'talk family'.

'Students reports', similar to other items, and 'information', related to 'talk family', were removed from subsequent analysis.

Only the item 'welcome', weakly negatively related to the subject 'discipline' coding, appears to have been influenced by these explanators in this subset. This relationship implies that the importance of a warm 'welcome' had been slightly greater to shipping students, which was to be expected, given that many of the logistics students had been recruited to study stage one of the course at other institutions.

Summary

Taken overall, the objective explanatory variables of 'years of experience', 'gender' and academic 'discipline' provided only limited explanation of the items which were most important to Masters level students in these particular decisions to undertake study in International Shipping and Logistics. In particular, 'gender' failed to provide a statistically significant relationship with any of the items and groupings shown above, possibly reflecting courses with limited total sample numbers of females. 'Length of experience' provided some weak statistical explanations of trends, but it was the difference between shipping and logistics 'disciplines' which furnished the greatest evidence of variations in the importance attached to different explanatory items. Even here however, there was little conclusive evidence to imply that for example a Masters level student of International Shipping was likely to display a particular set of objective traits. By implication, the proposition that an individual who displays a particular set of traits may be more likely to display an interest in studying International Shipping at Masters level is also likely to be flawed.

Explanations of the Distance Measures

Inter-correlations

To date, eight formulae have defined the set of distance measures required to represent differences between students' maps. Could this be reduced without significant loss of information? The empirical analysis proceeded to test this proposition by computing non-parametric measures of association between the different distance measures for each data set. Non-parametric measures were used, because much of the original data was found to be non-normally distributed (see Chapter 6). Using this approach, any redundant formulae or data should be revealed. Spearman's rho rank correlation coefficients were computed in order to compare the mean values of each of Langfield Smith and Wirth's (1992) distance measures. Mean scores for each individual were examined initially, but the analysis was then extended to cover full sets of pair-wise comparisons, as reported in Tables 7.7 to 7.12.

Inter-correlations between Mean Scores for Each Individual

In the 1997 data set (Table 7.7), several points are noteworthy, namely that:

1. High inter-correlations between F7 and F8 (rho = 0.96) and F6 and F8 (rho = 0.93) infer a high degree of redundancy between these measures in this data set. F4 was also related to F8 (rho = 0.72).
2. Measures F3 and F5 were related (rho = 0.85).
3. Dmax, the raw distance measure, was largely unrelated to the other measures.

In the 1998a data set (Table 7.7), it is apparent that:

1. Dmax was closely related to the mean of F5 and F6, both in turn related to each other.
2. Measures F3 and F4 were related (rho = 0.61).
3. Measures F7 and F8 displayed no systematic inter-correlations.

aying different aspects of these

:tween mean scores for each
)

and their significance (sig.)

29		
78		
48	0.90	
92	.000	
45	0.93	0.96
95	.000	.000

9	
0	
04	-0.02
7	.459

In the 1999 data sets (Table 7.8), it emerged that:

1. Measure F8 was closely correlated with F4 (rho = 0.96), F6 (rho = 0.98) and F7 (rho = 0.99).
2. Measures F3 and F5 were closely correlated (rho = 0.95).
3. Dmax showed some relationship to F3 (rho = 0.61) and F5 (rho = 0.49).

In reviewing the mean scores overall as compared between individuals:

1. Measures F7 and F8 were closely related, often also with the distance measures F4 and F6.
2. Measures F3 and F5 were usually related.
3. The raw distance measure F1 (Dmax) was less likely to be related to other variables.

The analysis then proceeded to examine the correlations between measures as computed for all pair-wise sets of observations.

Table 7.8 Spearman's rho correlations between mean scores for each individual (1999 data)

Rho coefficients between distance measures and their significance (sig.)

1999 data. Sample size = 54.

		F3	F4	dmax	F5	F6	F7
F4	rho	0.02					
	sig	.883					
dmax	rho	0.61	0.18				
	sig	.000	.186				
F5	rho	0.95	0.03	0.49			
	sig	.000	.853	.000			
F6	rho	0.01	0.94	0.17	0.01		
	sig	.930	.000	.218	.937		
F7	rho	0.06	0.95	0.23	0.05	0.98	
	sig	.672	.000	.092	.730	.000	
F8	rho	0.06	0.96	0.21	0.06	0.98	0.99
	sig	.650	.000	.130	.634	.000	.000

Source: the author

Table 7.9 Correlations between all pair-wise sets of observations (1997 and 1998a data)

Rho coefficients between distance measures and their significance (sig.)

1997 data. Sample size = 703.

		F3	F4	dmax	F5	F6	F7
F4	rho	0.68					
	sig	.000					
dmax	rho	0.22	0.17				
	sig	.000	.000				
F5	rho	0.80	0.43	0.31			
	sig	.000	.000	.000			
F6	rho	0.48	0.75	0.03	0.34		
	sig	.000	.000	.395	.000		
F7	rho	0.54	0.73	0.19	0.50	0.93	
	sig	.000	.000	.000	.000	.000	
F8	rho	0.55	0.74	0.17	0.49	0.94	0.96
	sig	.000	.000	.000	.000	.000	.000

1998a data. Sample size = 1081.

		F3	F4	dmax	F5	F6	F7
F4	rho	0.34					
	sig	.000					
dmax	rho	0.42	0.23				
	sig	.000	.000				
F5	rho	0.65	0.18	0.68			
	sig	.000	.000	.000			
F6	rho	0.22	0.85	0.32	0.27		
	sig	.000	.000	.000	.000		
F7	rho	0.23	0.86	0.35	0.29	0.99	
	sig	.000	.000	.000	.000	.000	
F8	rho	0.20	0.84	0.33	0.27	0.99	0.99
	sig	.000	.000	.000	.000	.000	.000

Source: the author

Inter-correlations between Pair-wise Sets of Observations

From the 1997 data set (Table 7.9), correlations revealed that:

1. Measure F8 was related to F6 (rho = 0.94) and F8 to F7 (rho = 0.96).
2. Measure F3 was related to F5 (rho = 0.80).
3. Dmax was not closely correlated with any measure.
4. Measure F4 was slightly more closely correlated with measures F6, F7 and F8, than F3.

In the 1998a aggregate data set (Table 7.9):

1. Measures F6, F7 and F8 were almost perfectly correlated (rho = 0.99) and F4 was related to all three (0.84 <= rho <= 0.86).
2. Measures F5 and F3 were correlated (rho = 0.65).
3. Dmax was most closely related to F5 (rho = 0.68).

In the 1999 data set (Table 7.10):

1. Measure F6 was correlated with F7 and F8 (rho = 0.99) as was F4 (rho = 0.90).
2. Measures F3 and F5 were closely related (rho = 0.88).
3. Dmax was not closely correlated with any other measure.

Table 7.10 Correlations between all pair-wise sets of observations (1999 data)

Rho coefficients between distance measures and their significance (sig.)

1999 data. Sample size = 1431.

		F3	F4	Dmax	F5	F6	F7
F4	rho	0.25					
	sig	.000					
Dmax	rho	0.42	0.25				
	sig	.000	.000				
F5	rho	0.88	0.23	0.44			
	sig	.000	.000	.000			
F6	rho	0.18	0.89	0.17	0.12		
	sig	.000	.000	.000	.000		
F7	rho	0.21	0.91	0.22	0.15	0.99	
	sig	.000	.000	.000	.000	.000	
F8	rho	0.21	0.90	0.21	0.16	0.99	0.99
	sig	.000	.000	.000	.000	.000	.000

Source: the author

Overall, analysis of pair-wise correlations suggests that:

1. Measure F4 was closely correlated with measures F6, F7 and F8.
2. Measures F3 and F5 were closely related.
3. The simple distance measure Dmax (Formula F1) was generally poorly related to other measures.
4. Future work might usefully concentrate on analysis of the raw distance measure (F1), a simple measure of differences (e.g. F3 or F4) and a more complex measure (e.g. F6 or F7).

On the basis of these known inter-correlations, variations in formulae were then examined using objective explanatory variables.

Aggregate Explanatory Models

Could distance measures be related to objective explanatory variables for the combined data set? For the combined 1997-1999 Masters data set (n = 139), four measures were analysed:

1. F1, the simple distance measure.
2. F3, a relatively simple distance measure.
3. F7, the more complex measure.
4. F71, taken as the square root of F7, to minimise any non-linear effects.

For the combined 1997-1999 aggregate data set, non-parametric correlations between measures (Table 7.11) indicated that only Formulae F1 and F3 were significantly correlated. Simple bivariate regressions of the measures against years of experience, gender and academic discipline yielded little useful information and coefficients are not reported in detail. Finally a stepwise procedure was used to regress the distance measures against a range of the key items identified above. The only explanatory variables individually significant in any circumstances were 'talk students', in relation to measures F3 and F7 and 'talk family' in relation to measure F3, and neither have any strong a priori logical relationship. Even for measure F3, where 12 explanatory variables explained only 21 per cent of its variation, aggregate explanatory analysis proved to be fruitless.

Table 7.11 Correlations between all distance measures (1997-1999 data)

Rho correlation coefficients between distance measures and their significance (sig.)
1997-1999 data. n = 139.

F3	rho	0.26		
	sig	.00		
F7	rho	-0.15	0.12	
	sig	.08	.16	
F71	rho	-0.15	0.12	1.00
	sig	.08	.16	.00
		F1	F3	F7

Regressions on objective explanatory variables 1997-1999 data. n = 139.
Level of significance p < 0.05 = *, P <0.01 = **, P < 0.001 = ***.

	Constant	Years of experience	Gender	Discipline	R^2
F1	***				.04
F3	***			**	.06 *
F7	***				.02
F71	***				.02

A stepwise procedure regressing distance measures against key items. 1997-1999 data. n = 139.

		Explanatory variables	
	R^2	Number	Significant variables
F1	0.12	11	No explanatory variables were significant
F3	0.21**	12	Talk family, talk students
F7	0.11	11	Talk students
F71	0.09	11	No explanatory variables were significant

Source: the author

Issue-based Explanations of Mean Distance Measures

In extending the attempt to build aggregate relationships between matrix distance measures, a combination of issue-based and objective explanatory variables were added to the analysis. However, all failed to provide statistically convincing results. Table 7.12 shows the results of stepwise multiple regression between all seven formulae, and explanatory variables. Even for measure F3, 83 per cent of the variation remained unexplained, and even though the influence of family and friends and pecuniary considerations were both positively related to F3, academic discipline (where shipping was coded as 0 and logistics as 1) was negatively related. For measures Dmax and F5, no statistically significant explanatory variables were found. On all other measures, the attractions of postgraduate study provided a significant positive link, but explained only 10 per cent of the variation in measures.

Table 7.12 Issues which 'explained' distance measures (1997-1999 data)

Data show regression coefficients, t values and coefficients of determination (R sq.) between measures and explanators (X1 to X3).

n = 139	Constant	X1	X2	X3	R.sq.
F3: Issue		Friends	Discipline	Money	0.17
Coeff.	1.068	0.131	-0.217	0.096	
t	12.108	2.838	-2.643	2.171	
F4: Issue		Postgrad			0.05
Coeff.	0.892	0.745			
t	1.561	2.624			
Dmax, F5	No variables entered				
F6: Issue		Postgrad			0.07
Coeff.	0.439	0.252			
t	2.751	3.173			
F7: Issue		Postgrad			0.09
Coeff.	0.724	0.612			
t	2.141	3.641			
F8: Issue		Postgrad			0.09
Coeff.	0.624	0.515			
t	2.170	3.604			

Source: the author

Disappointingly, the attempt to adequately fit aggregate functional relationships to the combined 1997-1999 data set was abandoned. Instead, the analysis then proceeded to apply stepwise multiple regression procedures to 'explain' distance measures in terms of the ratings of the main issues that had been identified as influencing the decision to study, for each individual data set. For each year, each of the distance measures was regressed on the main issues known to influence the study decision. The summary of the best-fit regressions, including up to three variables entered, are reported in Tables 7.13 and 7.14.

For the 1997 data set (Table 7.13), no suitable explanators were found for the simple distance measures, F1 or F3, although ratings of the importance of the reasons for undertaking postgraduate study in shipping or logistics were found to explain 33 per cent of the variation in measure F4. This issue was also significant in explaining variations in F6 and F8, recording a weak positive influence in both cases. The reasons for choosing to study in the UK were significantly and inversely related to measures F6 and F8.

Statistically significant but relatively weak explanations were found for all the distance measures in the 1998 data set (Table 7.13). Twenty per cent of the variation in Dmax was explained by the role of money as a barrier, and also the reasons for not studying at Plymouth, representing two negative aspects of this decision. Sources of information were significant explanators of the other simpler distance measures, F3 and F5, but over 85 per cent of variation was still unexplained. The money issue again provided significant explanations of the more complex measures, F6, F7 and F8, but almost 90 per cent of variation was unexplained.

For the 1999 data set (Table 7.14), a very poor level of statistical explanation was achieved, with only the simplest of measures, Dmax and F3 being influenced by the main issues involved. Family and friends explained less than ten per cent of the variation in measure F3, and negative influences of reasons for not studying at all, or not studying elsewhere and the sources of information explained almost 40 per cent of the variation in Dmax.

Table 7.13 Issues which 'explained' distance measures (1997-1998 data)

Coefficients, t values and R square of regressing measures on explanators

1997	Constant	X1	X2	Rsq.
F3, Dmax, F5	None entered			
F4: Issue		Postgrad		0.334
Coeff.	0.242	0.434		
t	2.157	4.249		
F6: Issue		Postgrad	UK	0.265
Coeff.	0.533	0.161	-0.376	
t	3.506	2.346	-2.733	
F7: Issue		UK		0.201
Coeff.	1.375	-0.773		
t	5.525	-3.064		
F8: Issue		Postgrad	UK	0.281
Coeff.	0.778	0.189	-0.517	
t	4.116	2.213	-3.026	
1998 F3:		Info		0.156
Coeff.	0.262	0.119		
t	3.442	2.880		
F4: Issue		Money		0.132
Coeff.	1.431	0.646		
t	3.948	2.614		
dmax		NotPlym	Money	0.198
Coeff.	21.63	3.796	4.664	
t	6.025	2.167	2.407	
F5: Issue		Info		0.128
Coeff.	0.322	0.113		
t	3.988	2.571		
F6: Issue		Money		0.109
Coeff.	1.029	0.533		
t	3.090	2.345		
F7: Issue		Money		0.104
Coeff.	2.202	1.160		
t	2.960	2.286		
F8: Issue		Money		0.106
Coeff.	1.734	0.795		
t	3.433	2.307		

Source: the author

Taken overall, these aggregate models of mean distance measures provide very poor explanations of the observed variation. For the combined 1997-1999 data, the attractions of postgraduate study provided the most consistent explanator of more complex measures. This was also important in the 1997 data along with the attractions of study in the UK. Financial barriers to study provided the most consistent explanator in the 1998 data, with information a secondary source. For the 1999 data set, no consistent variables featured as explanators, with no variables being entered on several of the distance measures. On the basis of these findings, any additional aggregate analysis of distance measures using issue-based explanators would appear to add little understanding of variation in the distance measures. The next step in the analysis used similar procedures to explain the pair-wise distance measures.

Table 7.14 Issues which 'explained' the distance measures (1999 data)

Data show regression coefficients, t values and coefficients of determination (R sq.) between measures and explanators (X1 to X3).

1999	Constant	X1	X2	X3	Rsq
F3: Issue		Friends			0.084
Coeff.	1.136	0.131			
t	10.848	2.185			
dmax: Issue		Notstudy	Info	NotElse	0.378
Coeff.	25.157	1.880	-1.919	2.425	
t	14.073	2.722	-2.105	3.416	

No issues were entered for measures F4 to F8 inclusive

Source: the author

Issue-based Explanations of Pair-wise Distance Measures

For the 1999 data, the largest single set, an attempt was made to 'explain' the individual pair-wise sets of formulae values measuring differences between maps, by regressing them on pair-wise differences in the perceived importance of each of the main issues in the study decision. Although interesting, the approach yielded weak empirical results (Table 7.15), with poor coefficients of determination, but individual variables recorded highly significant coefficients, due to the large sample size (n = 1431).

Table 7.15 Regression of individual pair-wise distance formulae values against differences in the perceptions of major issues (1999 data)

Data show regression coefficients, t values and coefficients of determination (R sq.) between measures and explanators (X1 to X5). Plym = Plymouth, NotElse = Not Else, Teach = Teaching.

1999, n = 1431	Significant issues and variables						
	Constant	X1	X2	X3	X4	X5	Rsq.
F3: Issue		Postgrad	NotStudy	UK	NotElse	Age	0.03
Coeff.	1.27	0.099	0.059	0.075	-0.129	0.088	
t	27.3	2.77	2.49	2.33	-5.08	1.99	
F4: Issue		Postgrad	Plym	City	Teach	Nation	0.06
Coeff.	1.92	0.494	0.384	-0.319	-0.171	-0.611	
t	13.8	4.43	4.22	-3.58	-2.11	-4.14	
dmax: Issue		Postgrad	Plym	Friend	NotElse	Teach	0.07
Coeff.	27.0	2.997	1.424	0.754	-1.159	-1.279	
t	50.2	7.54	4.32	2.65	-4.06	-4.48	
F5: Issue		NotStudy	UK	NotElse	Age		0.03
Coeff.	0.43	0.026	0.050	-0.040	0.040		
t	26.2	2.99	4.35	-4.38	2.50		
F6: Issue		Postgrad	Plym	Money	City	Nation	0.04
Coeff.	0.98	0.148	0.166	-0.090	-0.152	-0.231	
t	13.0	2.68	3.74	-2.44	-3.46	-3.13	
F7: Issue		Postgrad	Plym	City	Age	Nation	0.06
Coeff.	1.65	0.312	0.429	-0.028	0.337	-0.587	
t	11.7	2.86	4.92	-3.21	2.45	-4.05	
F8: Issue		Plym	NotPlym	Money	City	Nation	0.05
Coeff.	1.67	0.315	0.210	-0.172	-0.210	-0.418	
t	12.6	4.23	2.99	-2.75	-2.84	-3.34	

Source: the author

Results, reported so as to exclude variables with statistically significant coefficients beyond the fifth one in the analysis, revealed that:

1. For measures F6, F7 and F8, results are unstable, although the attractions of study at Plymouth feature highly in measures F6, F7 and

F8, as does the quality of life in the city. Reasons for postgraduate study and financial considerations are also important variables.
2. For measures F3 and F4, the attractions of postgraduate study present a common explanator, but no other variable features more than once.
3. Measures F3 and Dmax are related in that the attractions of postgraduate study and not seeking to study elsewhere feature in both, but other explanatory variables differ.
4. Measures F3 and F5 have four explanators in common, including reasons for not undertaking postgraduate study, the attractions of Plymouth and the UK and age.

Given these poor results the attempt to explain aggregate differences in maps by differences in the individual perceptions of each issue was not repeated for the other cohorts. However, an alternative approach, offering a single unified social cognitive theory of career and academic interest (Lent et al, 1994), remained to be explored.

Socio-cognitive Theories of Career Development

The shortcomings of inductive analysis of observed distance metrics are now apparent, and a new theoretical perspective, capable of generating testable hypotheses, is required to revitalise explanations of the observed empirical relationships. Lent et al's (1994) unified social cognitive theory of career and academic interest, choice and performance promised new explanations of why aspiring managers in shipping and logistics might choose to enrol on advanced courses. The approach is based on posits relating to the ways in which individuals manage issues of self-efficacy, expected outcomes from decisions and their personal goals. These issues are then postulated as being mediated by a range of personal factors, the contexts within which decisions are taken, and their previous experiences.

An immediate problem is encountered in calling on social cognitive theory to explain differences in cognitive maps that were developed under a different guise. The mapping studies were empirically driven and based on locally defined and contextually dependant concerns derived from focus groups. As such, some incongruence is inevitable between the issues expressed in focus groups, and the definitions of particular testable items within these issues, compared with the unified theory. This incongruence, coupled with differing primary research objectives dilutes attempts to test externally derived posits in the current context. However, to sufficiently replicate the contextual veracity or otherwise of such posits would require an externally imposed set of theoretical concerns, concepts and conditions, which would unacceptably suppress the richness of the local milieu. With these limitations in mind, the attempt to evaluate the explanatory contribution of Lent et al's approach within the current context proceeded.

Testable Hypotheses

Twelve predictions (H1... H12) each including a number of hypotheses were proposed by Lent et al (1994), but only those relating explicitly to academic interests which could be examined in relation to the available data bases are discussed below. Drawing on concepts derived from both educational and careers literatures, the overt actions of individuals were placed within the context of an interaction that influences the person-environment relationship. The behaviour of the individual was viewed within the context of a dynamic self, defining a triadic reciprocal causality. Some basic concepts include:

1. Self-efficacy, which involves each individual in judging his or her ability to achieve a set of outcomes. This state is dynamic, whereby as individuals gain additional evidence of past performance, they continually reappraise their perceptions of their capabilities.
2. Outcome expectations, which represent the effects of taking particular actions. These outcomes may relate to monetary, approval or self-satisfaction effects, and interact with self-efficacy.
3. Goals, which are in turn influenced by reflections on self-efficacy and expectations. Perceptions of such issues as employment conditions might logically be expected to influence postgraduate students.

In considering their career related behaviour, the background of an individual and their personal factors (gender, race / ethnicity, disability / health etc.) were hypothesised to influence their learning experiences. These learning experiences in turn will impact on the self-efficacy and the outcome expectations of the individual, with self-efficacy also influencing outcome expectations directly (H7).

At the centre of the model is a chain of links between career interest, intentions, activity and attainment. Career interest influenced career choice goals and intentions for activity involvement (H3), which in turn influenced choice actions, selection and practice (H4). Both of these links were also being moderately mediated by contextual influences proximal to choice behaviour. Finally, performance attainments, such as goal fulfilment and skill development followed on from choice actions (H5).

Self-efficacy and outcome expectations were both connected directly with components of this central chain of career interest / intentions / activity and attainment. Self-efficacy was postulated to influence directly the components of career interest (H1), choice goals (H10), choice actions (H11), and performance domains (H12). Outcome expectations were also posited to influence career interest (H2), choice goals (H8) and choice actions (H9).

In a final 'outer-loop', self-efficacy also influenced performance attainments (H12), which in turn influenced learning experiences (H6), before looping back to self-efficacy.

Respondent Profile and Methodology

Earlier in this chapter academic discipline was found to account for much of the variation between the maps of sample sub-groups. In order to neutralise this effect a subgroup of shipping students presented a distinct group on which to work with Lent et al's hypotheses. This group consisted of international shipping students on the Plymouth Masters course, including 21 from the 1997 cohort, 23 from January 1998 (1998a) and an additional student in the March 1998 survey (1998b). Characteristics of the samples selected are summarised in Table 7.16. Many of the respondents in this group were male, aged under 30, and represented many nationalities. Most claimed to have had at least some work experience, with significant proportions registering several years. Within the taxonomy presented, about one-third of students claimed to have gained mainly seagoing experience, although the loosely defined port and inland groupings may have also included broadly 'shipping' experience. Samples included all members enrolled in their particular cohorts.

each category (unless shown

1997	1998a/b
81	96
90	75
10	21
0	4
14	4
5	0
29	13
37	34
5	8
5	25
5	4
0	4
0	8
29	39
43	9
10	48
55	9
15	26
30	65
21	24

rised in Tables 7.17 and 7.18
each comprised several sub-
tiple components, drawing on
ence and hence any close

: normality in data sets. Hypotheses
es of causation between explanatory
gression.

ons of shipping Masters students

nts in each group of importance.
cision to study were:

nt	Important	Critical
	0	0
	35	13
	21	33
ipping at postgraduate level		
	0	5
	39	48
	33	42
ff studying shipping at		
	0	0
	18	26
	38	25
rses in these areas?		
	0	0
	48	18
	38	17
ipping in the UK attractive?		
	0	0
	52	26
	50	33

f shipping Masters students

:ach group of importance.
› study were:

Important	Critical
/ of shipping at Plymouth?	
5	0
39	26
29	33
0	0
26	21
37	17
tudy of shipping at	
0	0
35	4
42	8
0	0
26	13
33	13
tion etc?	
38	0
39	9
38	12
0	0
26	26
17	38

to that of the whole sample (see Chapter 5). Their reasons for choosing to study shipping at postgraduate level, particularly in the UK and at Plymouth were the most important issues in their overall decision, and, predictably, sources of information were an important issue. A larger than expected proportion, over 70 per cent, were influenced by family and friends and up to 30 per cent had not considered the downside of this particular course of action. Potential barriers to study had included teaching methods and financial considerations, ahead of quality of life issues.

Empirical Results

> Prediction 1. An individual's occupational or academic interests at any point in time are reflective of his or her concurrent self-efficacy beliefs and outcome expectations.
>
> Hypothesis 1A. There will be a positive relation between occupationally relevant self-efficacy beliefs and (expressed or inventorised) vocational interests.
>
> (Lent et al, 1994)

Few correlations linking self-efficacy statements (e.g. S54 and S57 in Table 7.19) with vocational interest (S15, S16, S40, S47) were significant (see Panayides and Dinwoodie, 1999, p.760). A non-significant link (1998b data, rho = 0.25) between those who valued their Plymouth alumni status (S54) and those studying in the language of shipping to gain more opportunities (S40) may imply that they were reinforcing an extant desire to gain further opportunities.

In similar vein, the only statistically significant link was between those who were attracted to Plymouth because it had accepted them (S57) and the opportunities offered by studying in the land of the language of shipping (S40 for 1997, rho = 0.44). A change in sign or strengthening in the correlations between 1998a and 1998b data on several items linked with S57 was noteworthy, where for example rho against S40 shifted from 0.16 (1998a) to 0.41 (1998b). By inference, the importance placed on having gained a place at Plymouth became more strongly related to vocational interests as the length of time spent there increased. The instances identified which supported Hypotheses 1A were neither consistent across all three data sets, nor across different sets of statements. At least in part, broad and often multi-faceted concepts that comprised the instrument employed here may have blurred some of these theoretical relationships.

Table 7.19 Correspondence between empirical statements and Lent et al's concepts

Key to Statements

Self-efficacy
S54: Studying shipping at Plymouth was attractive because I had already studied there.
S57: Studying shipping at Plymouth was attractive because they accepted me.
S71: I might have been put off study if Plymouth considered my grades to be inadequate.
S86: My company or parents partly funded and encouraged me.
Outcome expectations
S16: I studied at postgraduate level to specialise in shipping, as I want to work there.
S40: I studied shipping in the UK... the language of shipping, to gain more opportunities.
S47: I studied at Plymouth: its worldwide reputation is important when looking for a job.
Vocational Interest: S16, S40, S47, and
S15: I studied shipping to broaden my opportunities / be sure to find a job.
Interest S11
Choice action S75
Entry behaviour S27
Career / academic interest: S15 and
S14: I studied shipping to change my career / go ashore.
Academic outcome expectations
S21: I might have been put off study because I needed operational experience to improve my understanding.
S23: I might have been put off study as I had to be sure it was the right course.
S70: I might have been put off study at Plymouth if my qualifications were not considered unique by employers.

Source: the author

Hypothesis 1B. There will be a positive relation between occupationally relevant positive outcome expectations and (expressed or inventorised) vocational interests; negative outcome expectations will relate inversely to vocational interests.

Through having chosen to enrol on a Masters course in shipping, all study participants had indicated a substantial degree of vocational interest. The simple proportions of students to whom outcome expectations (e.g. S16, S40 in Table 7.20) were important thus provide some evidence of a link between the concepts. However, this excludes cases where negative outcome expectations relate inversely to vocational interests. Where these issues were considered to be 'important' to at least 42 per cent of students, and 'critical' to 30 per cent, and study as a postgraduate specialisation (S16) was 'relevant' to over 80 per cent of all groups surveyed, this evidence supports Hypothesis 1B.

Table 7.20 The importance of outcome expectations in the decision

Data shows the percentage of persons who agreed with the statement :

Year	Irrelevant	Relevant	Important	Critical
15: I studied shipping to broaden my career opportunities / be sure to find a job.				
1997	24	28	24	24
1998a	17	13	30	39
1998b	17	17	29	37
16: I studied shipping at postgraduate level to specialise in shipping, as I want to work there.				
1997	19	24	28	29
1998a	9	26	17	48
1998b	8	21	25	46
40: I studied shipping in the UK... the language of shipping, to gain more opportunities.				
1997	29	29	19	24
1998a	39	13	17	31
1998b	25	33	13	29

Source: the author

Hypothesis 1C. An additive combination of self-efficacy and positive outcome expectations will account for more variance in career / academic interests than will either self-efficacy or outcome beliefs alone.

In testing Hypothesis 1C, multiple regression was used to evaluate whether S14 and S15 (Table 7.21) as measures of the variance in career / academic interests were more dependent on an additive combination of self-efficacy (S54, S57, S71, S86) and positive academic outcome expectations (S21, S23, S70) than either in isolation. There was no evidence to reject this hypothesis, with 1997 data on variable S14, a career change variable, showing self-efficacy to be the sole source of variation, with no variation in 1998b data, and 1998a data supporting the hypothesis. Variable S15, relating to a desire for broader opportunities, supported the hypothesis in the 1997 data, with little variance apparent in 1998a/b data.

> Hypothesis 1E. A significant portion of variance in vocational interest stability will be accounted for by stability in self-efficacy and outcome expectations.

In testing Hypothesis 1E, S14 and S15 were regressed against the same set of self-efficacy variables, but with a different set of outcome expectation variables, (S16, S40, S47), related more specifically to interest in a career in shipping (Table 7.21). S14, the desire to change career variable, was more dependent on self-efficacy variables in the 1997 cohort, but more on the outcome expectation variables in both sets of 1998 data. The 1998a data set supported the hypothesis of a combined total variation in S14 that exceeded either the outcome expectation or self-efficacy components, and variation in S14 in the other two data sets equalled that of one of the other factors. Using variable S15, apart from 1998a data with little variation on these variables, one data set attributed all variation to self-efficacy and one to outcome expectations. Taken overall, there was again no evidence to deny the veracity of Hypothesis 1E.

> Hypothesis 3B. Occupationally relevant self-efficacy will relate positively to entry behaviours (e.g. information and job searches, applications for admission / employment, declarations of an academic major, attained choices).

From Table 7.22 it is apparent that a course specialisation in shipping was its principal attraction. For those who chose to do so, talking to students or former students presented an important item in their decision, with an absence of negative reports also being important to them. In the main talking to others who planned to join the course was less important, but literature sources were much more important to them.

Hypotheses 1C and 1E

ctation (OE) factors used to explain
in the final equation, after meeting a
ility of F>0.05.

es	R.sq. %	Factors
	25	SE, OE
	25	SE
	0	OE
	35	SE, OE
	0	SE
	18	OE
	0	SE, OE
	68	SE, OE
	43	SE
	59	OE
	0	SE, OE
	0	SE, OE
	25	SE, OE
	25	SE
	0	OE
	50	SE, OE
	0	SE
	38	OE
	23	SE, OE
	0	SE
	23	OE
	43	SE, OE
	43	SE

: decision to study shipping

atement:

	ortant	Critical
e specialises in shipping.		
	43	33
	52	26
	37	38
gazines / books / brochures.		
	24	10
	30	9
	21	8
:.		
	5	5
	0	4
	0	8
the course.		
	4	10
	7	22
	5	25
f students had given me bad		
	9	14
	0	4
	2	25

shows correlations between The best proxy indicators of le here included prior study ate variables providing some 'relevant' by 38 per cent. However in the succeeding cohort, S54 was rated as 'important' by only 8 per cent, but S27 by 65 per cent. In general, not a single statistically significant correlation was found between either of these sets of variables (Table 7.23). Given the small sample sizes and incongruence of the surrogate variables used to the concepts tested, this merely infers that in this particular study, no associations were found between occupationally relevant self-efficacy and entry behaviours. These results denied further testing of related hypotheses, where for example, Hypothesis 3D postulated that 'correlations between self-efficacy beliefs and entry behaviours will be reduced but not eliminated when the influences of vocational interests and goals are controlled'.

> Proposition 5. People will aspire to enter (i.e. develop choice goals for) occupations or academic fields that are consistent with their primary interest areas...

> Hypothesis 5A. There will be a positive relation between indices of (expressed or inventorised) interest and choice goals (e.g. aspirations, expressed choices).

In testing Lent et al's (1994) Hypothesis 5A, the percentage of Masters students enrolled on shipping courses, representing an expressed choice, agreeing with key statements, was high. Relevant statements included S16, relating to study of shipping in order to work there and S50 extolling a specialisation in shipping. Both were 'relevant' to over 80 per cent of students, the former 'important' to at least 60 per cent and the later at least 70 per cent (see Tables 7.20 and 7.22). The later appeared to strengthen with time elapsed since matriculation, becoming 'critical' for more students.

Consider elements of Lent et al's (1994) Hypothesis 10A, that:

> self-efficacy beliefs regarding particular career / academic activities will be positively related to the perceived amount of (a) personal success experiences; (b) exposure to successful models; (c) favourable social persuasory communications...

In terms of personal success experiences, variables S17 ('I studied shipping because I wanted more study following my work experience'), and S102 (years of experience) were used. Correlations with S54 and S57 as loose indicators of self-efficacy beliefs provided no statistically significant relationships.

The surrogates used to depict 'exposure to successful models' included talking to family or friends in the industry (S28) and wanting to study because friends had completed the course (S53). A non-significant link between S53 and S54 (relating to prior study at Plymouth, rho = 0.33 in 1998, 0.37 in 1998b) may furnish weak evidence of support.

Table 7.23 Correlation coefficients between self-efficacy and entry behaviour

Statements

S54: Studying shipping at Plymouth was attractive because I had already studied there.
S57: Studying shipping at Plymouth was attractive because they accepted me.
S27: Finding out about courses was important to me.
S31: I found out about courses by reading magazines / books / brochures.
S33: I found out about courses by talking to others planning to study there.
S34: I found out about courses by talking to students on the course.

Spearman's rho correlation coefficients

Statements	1997	1998a	1998b	Significance
S57: S27	-	-0.18	-0.37	
S54: S27	-	0.21	-0.06	
S54: S31	-0.06	-0.09	-0.16	
S54: S33	0.12	-0.05	0.02	
S54: S34	-0.13	-0.01	-0.04	
S57: S33	0.15	-0.17	0.09	
S57: S34	-0.25	-0.27	-0.07	

Source: the author

Finally, favourable social persuasory communications might have involved talking to others about the study decision including family and friends (S8), former lecturers (S29), others planning to study (S33) and enrolled students (S34). Again, even the only statistically significant link (observed in the 1998 data sets) between being attracted through prior study at Plymouth and talking to former lecturers (S29: S54, rho = 0.69 in 1998a and 0.42 in 1998b), could be explained by factors such as inertia.

Conclusions

The major problem encountered in testing Lent et al's (1994) useful hypotheses related to the incongruence of concepts, some of which were not uniquely represented by data from the Plymouth database, based on verbatim statements. In particular, measures of self-efficacy were often either oblique or linked with other concepts, and an absence of precise measures of other variables prevented testing of propositions 2, 4, 6, 7, 8, 9, 11 and 12.

Table 7.24 summarises some of the findings. In general, weak support was found for elements of propositions 1, 5 and 10. If an attempt were made to replicate similar studies at postgraduate level, problems might relate to:

1. Applying theories to mid career returners to study. Lent et al's (1994) work was aimed primarily at younger groups.
2. The theoretical emphasis placed on self-efficacy. Focus groups identified the issue, but did not reveal the same degree of emphasis as placed on it by Lent et al amongst these particular students.
3. Emphasis on perceived barriers to career, due to ethnicity and gender may be useful, in an international environment. Such factors have usually been studied at earlier career stages (e.g. McWhirter, 1997).
4. The mapping format adopted did not easily sit astride explicitly social cognitive theories, and is probably best viewed as a complementary approach based on differing objectives.

In concluding, definitive 'explanations' of the maps which students revealed are inappropriate. Indeed where a map is unique to one individual, even an attempt to reduce the maps of a few individuals to a standard format may devalue them, let alone any comparative or explanatory process. The worth of such attempts may lie in failure, where the uniqueness of each individual decision and an inability to categorise the diversity and variety of groupings of them, retains their wealth. The rich tapestry woven within each individual, expressed by them, as their account of their situation, suffices. There is a danger that researchers probing far beyond simple mapping and qualitative analysis of individual study decisions might tarnish this tapestry, generating spurious findings.

Table 7.24 A summary of the testing of Lent et al's hypotheses

Hypothesis	Comment
1A	Some weak support. The link between occupational self-efficacy and vocational interest appears to strengthen with time elapsed since matriculation.
1B	Evidence generally supported this.
1C	Two instances support this with no contrary evidence.
1E	One case supported this with no contrary evidence.
3B	No supporting evidence.
5A	Simple descriptive statistical support.
10A	No supporting evidence.

Source: the author

On this note of caution, it is time to report on a comparative analysis of national surveys of undergraduates and postgraduates studying similar academic disciplines.

8 A National Comparison of the Perceptions of Undergraduates and Postgraduates

Introduction

In this chapter we aim to extend the discussion to other educational institutions, in an attempt to draw more general findings. Much of the work reported to date has concentrated on one university, but if a wider applicability can be established, it may help to support a case for policy changes in national recruitment and the development of human resources in organisations engaged in international shipping and logistics. In particular, it is proposed to concentrate on key issues relating to the sources of information used by potential applicants, the role of family and friends, and more basic motives for study by potential recruits into the logistics and maritime industries. These issues, discussed in relation to the decision to study ISL at university, include some results from two surveys conducted nationally within the UK.

One of the surveys undertaken involved additional applications of the instrument described above, which was administered to postgraduate students of ISL at five other universities in the UK. In a second study, an instrument was developed to investigate aspects of the decisions of undergraduates to study maritime business at university, based initially on results from unstructured questionnaires, which were later refined into a tick-box instrument and piloted. The amended instrument was also administered nationally in the UK and the overall results are reported elsewhere (Dinwoodie, 2000). Some of the issues and items which emerged as being of concern to undergraduates in their study decision, also represented concepts which were sufficiently similar to concerns identified by postgraduates, enabling some comparison between perceptions of the two groups. These are reported in relation to the sources of information, the role of family and friends and more basic motives involved in making decisions to undertake advanced study in ISL.

Methodology and Sample Characteristics

Although the perception ratings scales that were employed differed between the undergraduate and postgraduate surveys, both involved a four-point scale (Table 8.1). Based on the assumption that these scales were broadly comparable, Pearson chi-squared tests of the degree of association between frequency distributions showing the percentage of the ratings for the two groups in each response category were conducted and are reported where appropriate. The chi-squared statistic is shown, along with (in parentheses) the level of significance, with which a null hypothesis of no association between the two distributions is rejected (Tables 8.2 to 8.5).

Table 8.1 Comparison of rating scales used in undergraduate and postgraduate surveys

Question: *How important was each reason / source etc.?*

Prompt to undergraduates	Prompt to postgraduates	Scale score
Very important	Critical	3
Quite important	Important	2
Indifferent	Relevant	1
Not at all	Irrelevant	0

Source: the author

From Table 8.2 it is apparent that almost one in four of both sets of respondents are female, yielding statistically similar proportions at both undergraduate and postgraduate levels. As might be expected, the age profiles of the groups are significantly different, with one-third of the 418 undergraduates sampled aged under 20 years. Nine per cent of undergraduates and 21 per cent of postgraduates are over 30 years of age.

In the combined samples, UK nationals presented a larger proportion of the undergraduates sampled than the postgraduates. This may reflect the sampling of more explicitly international postgraduate courses, and may not present a representative national picture of the entire population of UK students of shipping or logistics, which may include many more domestic students registered for non-international course options. Although non-Europeans comprised 16 per cent of the postgraduate sample, the

parent at this level, and also
ually, this trend towards the
y to reflect future trends in the

of the sampled groups

and chi square values

duates	Chi square
	0.01 (.9270)
	101.77 (.0000)
	117.70 (.0000)
rience	

and three-quarters of the
relevant work experience.

Sources of Information

Two differences are apparent between the prompts proffered to undergraduates and postgraduates, in relation to the sources of information, which influenced their decisions to study (Table 8.3). Firstly, undergraduates were given a normative prompt, contrasting with the single historic event offered to postgraduates. Secondly, the prompt given to undergraduates referred to 'courses and careers', whilst that to postgraduates related solely to 'courses'. Consequently, undergraduate responses also included more careers related issues, namely:

1. Talk to careers advisers.
2. Read job advertisements.
3. Write to companies direct.
4. Attend careers presentations.

Other postgraduate responses relating to the 'British Council', 'chance' and 'talking to' current, prospective or former 'students' were also not directly comparable.

Although there was no statistically similar distribution on any of the four measures of ratings of sources of information between undergraduates and postgraduates, some useful insights are apparent into the policy implications of the rating of each item.

'Talking to people in industry' was a more important source of information for undergraduates being 'quite important' to 65 per cent, than postgraduates (39 per cent), and 'very important' to almost half of undergraduates. The implication for increasing the opportunities for school and college leavers to have as much contact as possible with industries engaged in ISL is obvious.

'Talking to lecturers' was 'quite important' to 61 per cent of undergraduates but only 30 per cent of postgraduates, and again 'very important' to more undergraduates. The need to ensure that lecturers are *au fait* with current industrial developments must present an ongoing priority, if students are to receive accurate information.

Literature sources of information were important to two-thirds of undergraduates, but fewer recorded this item as 'very important' compared with 'talking to people in industry'. Five per cent fewer postgraduates regarded this source as important compared with 'talking to people in industry', again showing that whilst literature is an important source of information, the oral contact with practitioners is rated even more highly by potential applicants.

Table 8.3 Comparative ratings of the importance of sources of information

Data shows the percentage of item responses and chi-square statistic.
Undergraduates (Ug.): *How would you find out about courses and careers in maritime business? How important is each source?*
Postgraduates: (Pg.): *How did you find out about courses in these areas? How important was each source?*

Rating	Ug.	Pg.	Chi-square
Ug. response: *I would talk to people in industry.*			
Pg. response: *I talked to a work friend / people in industry.*			
Not important	30	41	88.02 (.0000)
Indifferent	5	20	
Quite important	19	26	
Very important	46	13	
Ug. response: *I would talk to my lecturers.*			
Pg. response: *My previous lecturers told me about it.*			
Not important	32	57	58.95 (.0000)
Indifferent	7	13	
Quite important	33	20	
Very important	28	10	
Ug. response: *I would read magazines, books, brochures.*			
Pg. response: *I read magazines, books, brochures.*			
Not important	19	42	66.72 (.0000)
Indifferent	15	24	
Quite important	32	22	
Very important	34	12	
Ug. response: *I would talk to friends / relatives.*			
Pg. question / response: *How did your family or friends influence your decision to study? How important were they? ...I talked to them.*			
Not important	37	32	18.78 (.0003)
Indifferent	21	28	
Quite important	23	31	
Very important	19	9	

Source: the author

'Talking to family and friends' was at least 'quite important' to relatively fewer undergraduates than other sources. However, at around 40 per cent of postgraduates it represented a higher proportion than for other sources. Taken overall, some 40 per cent of both undergraduates and postgraduates stated that they had been influenced by significant others in their decision to study ISL. This implies that such issues as a family tradition of seagoing and peer influences remain as ingrained influences, even on the composition of university classrooms today. If genuinely new blood is to be attracted, there is a need to raise the profile of ISL beyond the traditional groups who may have been employed in these industries.

The Perceived Importance of Self, Family and Friends

Knowledge of which individuals claim to own the decision to undertake advanced study is important to human resources managers and academic course managers, if they are to attempt to lure personnel to upgrade their skills and knowledge. In an attempt to address this issue, comparisons were made between undergraduate and postgraduate perceptions of the relative importance of:

1. Previous work experience.
2. Explicit attempts to broaden the range of employment opportunities open to them.
3. Parental and other pressures.
4. Peer pressure.

Not all items in this section are necessarily directly comparable, but the verbatim content of relevant prompts is also presented below for convenience (Table 8.4).

Given that a higher proportion of postgraduates had registered at least some relevant work experience, it is unsurprising that a higher proportion of them were influenced by, 'wanting more study following their previous work experience'. This issue was at least 'quite important' to 37 per cent of postgraduates compared with 28 per cent of undergraduates, although a higher proportion of the latter again claimed that it was very important to them. The implications of this finding are again obvious, in that course marketing materials, particularly for postgraduate courses, need to emphasise the attraction of courses to experienced individuals in helping them to contextualise and build on their prior experiences. Although

Table 8.4 Comparative ratings of the importance of self, family and friends

Data shows the percentage of item responses and chi-square statistic.
Undergraduates (Ug.): *What reasons made studying Maritime Business at university attractive? How important was each reason?*
Postgraduates (Pg.): *What reasons made studying Logistics / Shipping at postgraduate level attractive? How important was each source?*

Rating	Ug.	Pg.	Chi-square
Ug.: *I was attracted to study ... because of my previous work experience.* Pg.: *I wanted more study following my previous work experience.*			
Not important	61	44	45.91 (.0000)
Indifferent	11	19	
Quite important	12	29	
Very important	16	8	
Ug.: *I was attracted to study in this area because I needed to improve my job prospects / begin a career.* Pg.: *I wanted to broaden my opportunities / be sure to find a job.*			
Not important	34	13	49.82 (.0000)
Indifferent	13	22	
Quite important	20	36	
Very important	33	29	
Ug.: *I was attracted to study in this area because of parental pressure.* Pg.: *How did your family or friends influence your decision to study? How important were they? ...I talked to them.*			
Not important	77	67	16.25 (.0010)
Indifferent	12	12	
Quite important	7	9	
Very important	4	12	
Ug.: *I was attracted to study... because my friends were doing the same.* Pg.: *What reasons made studying logistics / shipping at Plymouth attractive? ...I have friends who have completed the course [there].*			
Not important	87	77	13.69 (.0034)
Indifferent	9	13	
Quite important	2	7	
Very important	2	3	

Source: the author

this may not be an issue for inexperienced individuals, its importance in luring experienced individuals from remunerated employment into further study may be substantial.

Whilst the influence of prior work experience on shaping the future of an individual must surely be something they can claim full personal responsibility for, perceptions of a need to undertake study because of 'a need to improve employment prospects' or 'begin a career' could be influenced by outside pressures. Parents, careers staff and teachers, peers or casual industrial contacts could all potentially influence an individual in perceiving a need to improve their long term employment prospects, or even embark on a particular career. In this context, this was an important issue to 65 per cent of postgraduates and 53 per cent of undergraduates, being 'very important' to at about one-third of both groups. This issue influenced all but a few postgraduates. Whatever the relative attractions and weightings given to subject or academic interest, and basic employment motivations as alternative reasons for seeking to undertaking advanced study, the latter is at least an issue for virtually all postgraduate vocational students and many undergraduates. The potential career openings available to successful graduates, the monetary and other advantages associated with those openings, and first hand accounts of such alumni are all potentially important influences on attracting individuals into particular courses of study. It is essential that this information is explicit in course marketing materials. It might also be useful for ongoing careers advice to be available to students whilst enrolled on courses, to help to focus and develop their exit strategies, in seeking to look ahead to their next career moves.

By way of contrast, relatively few students were willing to claim that 'significant others', including either their parents, or their peers had had major influences on their decision to embark on advanced courses of study. Surprisingly, 21 per cent of postgraduates, but only 11 per cent of undergraduates claimed that parents had had a significant influence on them. For 12 per cent of postgraduates, this had been a 'very important' influence, compared with only four per cent of undergraduates. In particular, family traditions or even funding for studies could have been involved in discussions with parents, but they were potentially significant to those individuals involved. Any unstated or implicit influences of parents and friends are not apparent from these statistics, nor are any attempts by relatively insecure new undergraduates to inflate their newly found independence, whilst away from home for the first time. For at least a significant minority of students, the yore relating to conditions, prospects and first hand experiences of existing employees in an industry have proved to be an important influence on their own choices. Whether real or otherwise, perceptions of the claim to independence are very high in relation to the role of peers on influencing study decisions. Only four per cent of undergraduates and ten per cent of postgraduates claimed to be influenced by what their friends had done. However, these influences may represent the tip of an iceberg in that a single negative report of a bad experience could have wide-reaching effects, and many more subtle and implicit influences are also surely at work.

In terms of marketing and publicity for courses, it is important that potential applicants can feel that they are personally responsible for making decisions regarding their own future, and it is likely that basic employment motives will be important in that decision, for virtually all postgraduates.

The Perceived Importance of Basic Motives in the Decision to Study

Just how important are basic motives in the decisions of different groups to undertake study in ISL? The attractions of 'credentialism' (Hesketh and Knight, 1999, p.152) were undeniable to both groups in this study. Table 8.5 shows that the attraction of study at a university with a 'worldwide reputation' was important for 57 per cent of undergraduates and 39 per cent of postgraduates. For the latter group, the basic motivation linking such a reputation with perceptions of future employment prospects was also explicit. Not only does a university need to ensure that it has a genuine worldwide reputation, but also it must be perceived as having such, and such perceptions need to be reinforced with good employment records in their alumni.

A potential deterrent to 66 per cent of undergraduates and 37 per cent of postgraduates could have arisen if they had 'found that my university's qualifications were not considered unique by employers'. This was not an issue for almost 40 per cent of postgraduates, probably because many of them had already established at least an initial employment track record, and were less concerned by the impacts of one year of study on their career tracks. The need for academics to ensure that close links with industry are maintained on vocational courses in ISL is one implication of this finding. Others include a need to ensure that contacts between students and potential employers maintain and enhance any existing impressions which may have been established, and in turn, to ensure that the quality of both courses and students on them are maintained at the highest levels possible.

Particularly for potential undergraduate applicants, the first destination records of course alumni are a significant factor in their choice of subject, course and university.

Table 8.5 Comparisons of ratings of the importance of basic motives

Data shows the percentage of item responses and chi-square statistic.

	Undergraduate (Ug.)	Postgraduate (Pg.)	Chi-square
Ug.: *What reasons attracted you to choose your present university ...to study Maritime Business? ...My university has a worldwide reputation.* Pg.: *What reasons made studying logistics / shipping at your university attractive? ...My university's reputation and tradition is worldwide, important when looking for a job.*			
Not important	30	37	33.41 (.0000)
Indifferent	13	24	
Quite important	29	28	
Very important	28	11	
Ug.: *What reasons attracted you to choose your present university ... to study Maritime Business? ... [It] offers a course which is good / unique.* Pg.: *What reasons might have put you off studying logistics / shipping at your university? ... If I had found that my university's qualifications were not considered unique by employers.*			
Not important	29	39	83.40 (.0000)
Indifferent	5	23	
Quite important	30	26	
Very important	36	11	
Ug.: *What reasons might have put you off studying Maritime Business at your present university? ... If I had been offered a job before arriving at university.* Pg.: *What reasons might have put you off studying logistics / shipping at postgraduate level? ... If I was offered a job giving good experience.*			
Not important	62	52	9.03 (.0289)
Indifferent	14	17	
Quite important	16	23	
Very important	9	7	

Source: the author

Although not widely rated as 'very important' in their study decisions, 25 per cent of undergraduates and 30 per cent of postgraduates said that they could have been deterred from study by 'an offer of employment affording good industrial experience'. This was not an issue to over 50 per cent of both groups, which implies a high degree of commitment to a decision to study once it has been made. However, these perceptions are reported *post hoc*, and it is possible that prior perceptions of potential barriers to study could have been much greater. This rating again emphasises the importance of employment motivations, at least to significant groups of potential students. If possible, larger proportions of students need to be attracted to courses as a first choice, and not merely because they were unable to find suitable employment. Increasingly, the challenge is to academics to be able to devise and deliver courses that not only simulate work experiences, but which also appear more attractive to potential applicants than an equal period of time spent in the workplace.

9 Attitudes, Education, State Controls and Practitioners in UK Port Logistics: a Soft Systems Approach

Introduction

Management education in shipping and logistics does not end with a graduation ceremony. Rather, this marks the beginning of a new career phase in which a practitioner must juggle the competing and continuing demands of the workplace against an increasingly pressing need to update personal professional skills and knowledge. In this chapter, we present a survey of the attitudes of a range of shipping and logistics practitioners towards UK port state control (PSC), as empirical evidence of the need for ongoing management education. The objective of this chapter is to apply a soft systems approach to investigate the attitudes of these groups towards aspects of PSC, achieved by surveying interested parties regarding their views of the role and effectiveness of PSC. The implications of these findings are then used to devise an educational strategy for practitioners, including inspectors and operators, to raise the effectiveness of PSC.

Inadequate implementation and insufficient enforcement of international conventions probably account for many of the substandard operators, ships and crews which are present in European Union (EU) waters (Dixjhoorn, 1993). Any substandard vessels present both a continuous threat for the marine environment and maritime safety, and also denigrate the reputation of the wider shipping industry. Press coverage of accidents at sea affecting EU coastlines has generated concern, and ongoing considerations for enhancing maritime safety and pollution prevention. However, a gulf exists between improved safety regulations, and practical rises in standards (Payoyo, 1994), with a need for controls to ensure that legislation is complied with. In recent years, co-operation in enforcement by port States resulted in regional PSC systems, including the Paris Memorandum of Understanding (MOU), signed by 14 European States in January 1982, designed to eliminate substandard ships from European waters and reduce accidents. Since then, similar agreements have been reached in other regions of the world, but their operational effectiveness may be raised if their managers are aware of the attitudes of practitioners towards such PSC systems. This chapter reports on such a study within the UK following a brief review of some perspectives on maritime safety, elements of a soft systems methodology, and a summary of the survey methodology. Some of the findings are reported before highlighting areas in which a strategy for improved management education and training could raise standards.

Some Practitioner Perspectives on Maritime Safety

The issues of marine safety and pollution are matters of international concern, with seaworthiness, safety and the protection of the marine environment of political interest. Safety, to ensure that human lives are not lost and that seas remain unpolluted, is embedded in regulations drafted by international and national authorities seeking to promote uniform standards, formulated by the International Maritime Organisation (IMO). Many of IMO's conventions and treaties have gained wide acceptance, where most shipping nations have ratified SOLAS (Safety of Life at Sea) and MARPOL (Marine Pollution convention). The most important treaties now cover much of the world fleet, and these internationally developed standards now form the basis of most national maritime legislation. The questions posed in the survey of practitioners reported below were devised around an informal application of systems concepts. In this approach, representing a measure of the perceived system performance, respondents including shipowners, classification societies, insurers, and port authorities were asked whether maritime organisations were perceived to be doing their job properly. System components and objectives were also investigated in relation to:

1. Which types of ships were perceived as being substandard.
2. What the role of PSC should be.
3. How the powers and procedures of inspectors could be enhanced.
4. Other changes which might be appropriate.

Soft Systems

'The role of PSC' is not a clearly defined prescriptive set of 'hard' actions that if enacted correctly, would result in the successful performance of a particular function. Rather, it involves a set of 'soft' ill-defined relations between human participants who share a common concern about a problem situation. In a soft systems approach, before deciding 'how' to intervene, those individuals or groups sharing a common interest in the situation would first seek to reflect on what the real problem at issue is. Actors in a complex human activity system (Wilson, 1990), interact with each other in the pursuit of a common purpose, but any recommendations as to how change may be implemented must follow this initial reflective process.

The port inspection system of interest in this study is limited to the UK, but the soft question of what the inspection process really involves, and hence any scope for change, remains. Training provision for inspectors is viewed as part of a human resources system, within which a desire to raise their effectiveness is an input into a transformation process enabling their development and which achieves an output of increasing their effectiveness (Figure 9.1). Logically, each entity within the human activity system must be dependent on other verbs or activities within it. The necessary training resources, including both human resources sufficiently skilled to provide training and technical information must be available if required. Measures of whether training has been successfully provided must be defined, which in turn are relevant to the fundamental purposes or objectives of the system of training provision. The continuing satisfactory achievement of performance is monitored on an ongoing basis, and if necessary, a series of control mechanisms may be applied to ensure that standards are maintained. Within the boundaries of their authority, managers of the system in question will require resources to ensure that it remains effective.

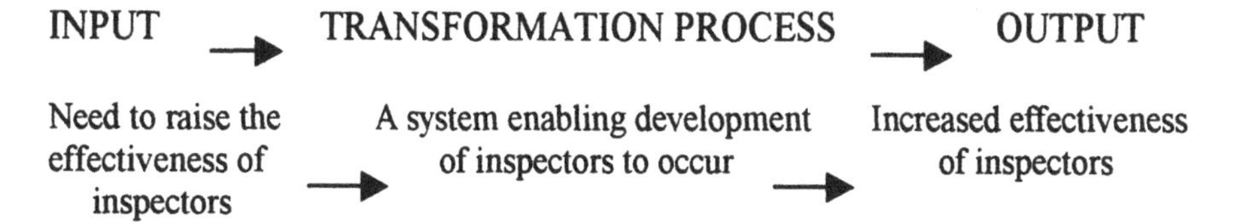

Figure 9.1 A definition of systems enabling education of inspectors to occur

Source: the author

The main stages involved in the soft-systems methodology (Wilson, 1990), are outlined in Figure 9.2. This commences with a variety of unstructured data collection procedures including discussions, questionnaires, interviews and other methods, which attempt to define and clarify the nature and linking of the issues which define an initial expression of concern. In structuring this complexity, each issue in a 'rich picture' as defined in the real world, is re-formulated in an abstract systems world, using a 'root definition'. Just as each participant in the system requires the system to achieve a particular outcome from their particular perspective, in turn defining what the system is from their stance, so each requires a different root definition, with increasing detail at higher levels of resolution.

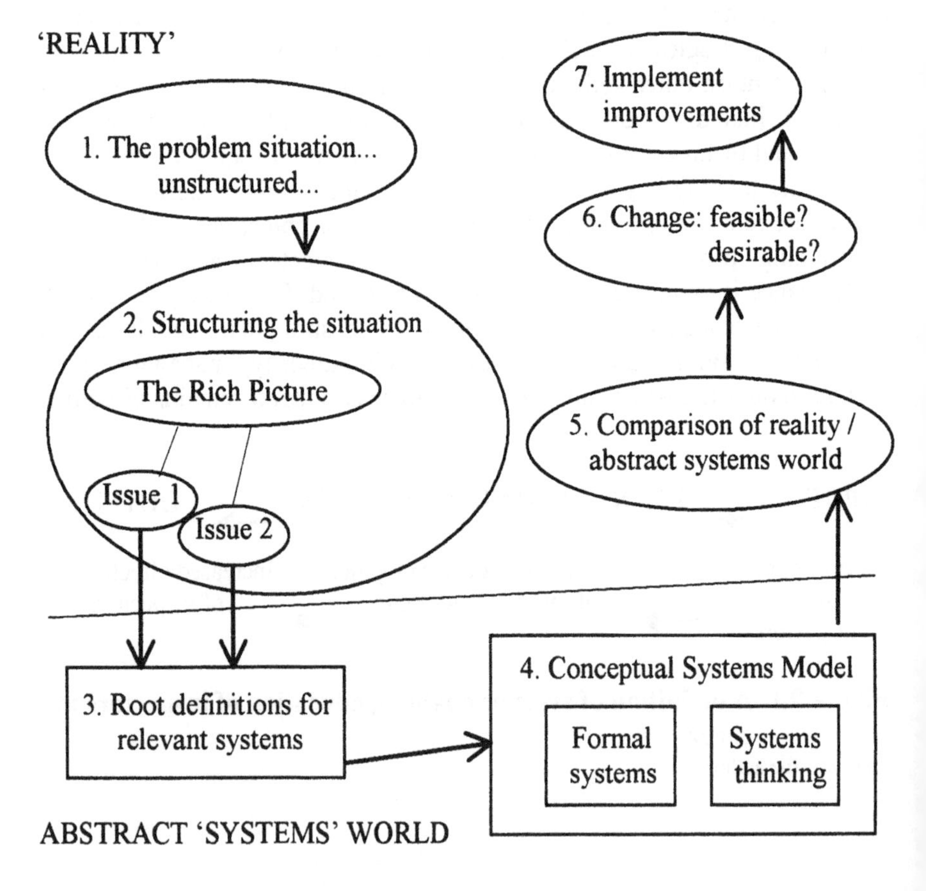

Figure 9.2 The soft systems methodology
Source: adapted from Wilson (1990, p.69)

In the systems world, the root definition contains the suggested set of actions, or transformations, which may alleviate the initial concerns of a problem owner. This idealised set of activities that the system must perform, representing a view of what exists from a particular stance, is termed a 'conceptual systems model'. Following this reflective process a comparison is then made between the rich picture and the conceptual systems model, before the feasibility and desirability of possible changes are discussed. Following this, an iterative process of implementing improvements, and if necessary redefining the nature of the problem owner's remaining concerns ensues.

A first level analysis of PSC might relate to its role in marine safety with the next level defining issues of education and training, penalties, frequency of inspection and the like, such as the rich picture in Figure 9.3. At level three, interest is centred on training and educational provision with a root definition and CSM from the inspector's stance (Figure 9.4).

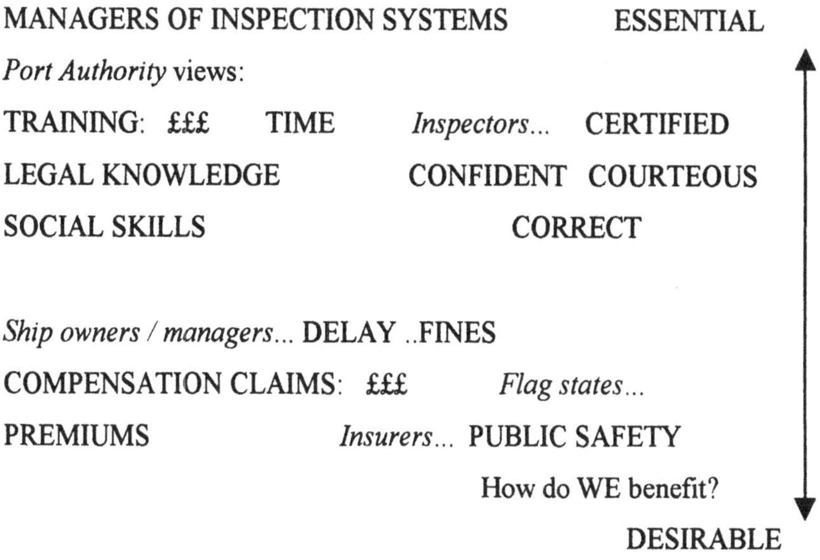

Figure 9.3 A rich picture of the training requirements for PSC systems

Source: the author

A root definition of training, viewed by managers of inspection services might be:

> a system to ensure that inspectors possess the knowledge and technical skills required to inspect vessels fairly and efficiently.

Root definitions from the port authority's stance may include:

Customer - the managers of inspection services.
Actor - the port authorities who fund training or conduct assessment.
Transformation - the process of acquiring knowledge and technical skills.
World-view - the person participating in training.
Ownership - the training provider.
Environmental constraints - the acceptability of qualifications and the training budget.

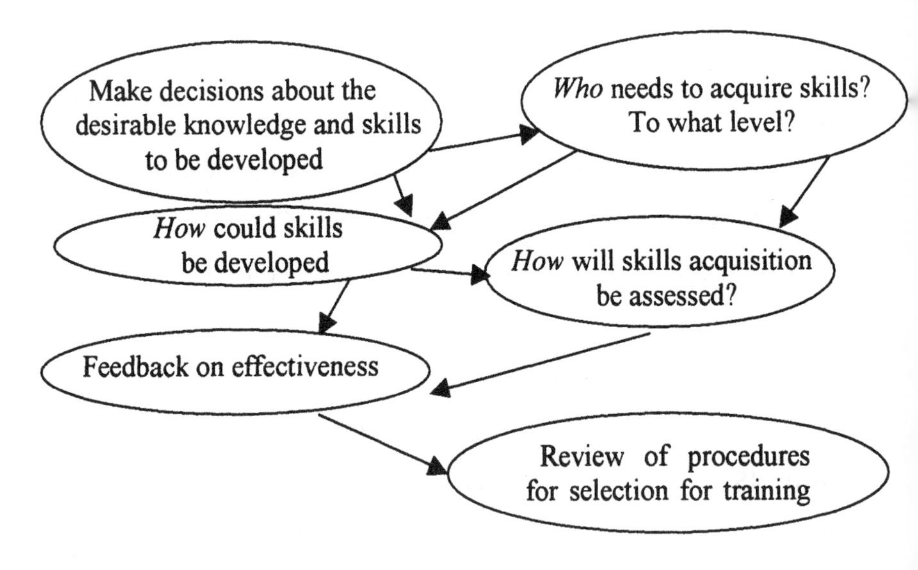

Figure 9.4 A Conceptual Systems Model of training provision for inspectors
Source: the author

However, despite the efforts of regulatory bodies, safety considerations may be subsumed by commercial pressures for some groups of shipowners, operators, managers, classification societies, insurers, flag States, notably third world shipowners, and Flag of Convenience (FOC) registries. Flag States, responsible for inspecting ships to ensure compliance with international requirements, play a crucial role in maintaining maritime safety standards, but concern has mounted where they may lack the resources or infrastructure required to monitor the international tonnage they so readily accept.

Traditional safety controls, imposed by shipowners, may become ineffective as professional ship management companies increasingly separate ownership from management in the shipping industry. Competition, inducing cost cutting measures, may impact on safety, and economic recession may delay decisions to purchase new tonnage, due to low returns on investment, and high capital costs and ageing ships are more likely to become unsafe, especially if inadequately maintained. Shipping related accidents cause fatalities, pollution and generate a poor image with the public and policy makers.

Pursuant on marine accidents, authorities lay down more regulations and recommendations, such as those by Lord Donaldson (1994). However, although the European shipping industry seems to agree that substandard shipping needs to be eradicated opinion on how responsible operations can be assured differs. At least in theory, education and training undoubtedly has a role to play (Dinwoodie, 1999a) where the classroom can provide a relatively risk free environment within which cognitive and affective change in operators and managers can be effected. At an enforcement level, the ability of IMO to ensure that oceans are safe and free from pollution has been questioned (Dijxhoorn, 1993). Despite the existence of legislation it is not uniformly enforced worldwide, as IMO has no enforcement role, and MOU's may differ in their approach to adopting IMO resolutions (IIJNET, 1998a). PSC as carried out in the USA by the Coastguard, established in Europe since 1982, and more recently in the Asia-Pacific, Latin America, Caribbean and Mediterranean regions, promotes enforcement, but is it the best solution? The cases for reviewing the role of classification societies (Boisson, 1994) and improved training of PSC inspectors (Clinton-Davis, 1994) are well voiced. However, a survey of UK maritime professionals regarding the role of PSC in raising ship safety standards and its implications for developing education programmes may suggest other approaches.

The Survey Methodology

Questionnaires were sent to 200 shipping organisations in the UK. Of 57 replies received, 23 were from port authorities, 20 from shipowners and managers, 6 from maritime organisations, and 4 each from consultants / surveyors, and insurers / P&I Clubs. No classification society responded. The sample collected should be viewed as indicative rather than exhaustive or even necessarily representative, but to avoid the problems of small samples, statistical information is only reported for owners and port authorities. The survey, conducted contemporaneously with the release of the Donaldson (1994) report, aimed to determine views on recent system performance measures relating to whether various maritime organisations were felt to be carrying out their duties properly, or whether there were felt to be too many substandard ships. In terms of reviewing or reformulating objectives, should authorities concentrate on drafting more regulations, or enforcing them, and in terms of system purpose and domain, did PSC have an important role to play, and had it been successful in Europe and the USA? Issues of system boundaries, spatial and otherwise, transferability, holism and information requirements were raised by asking whether PSC should have enhanced powers, be extended to other regions, and was more co-operation and the establishment of a large database needed? Systemic concepts of boundaries and equifinality were invoked by asking whether PSC could be used as a trade barrier, and might not port States be infringing on doing the work of others? Questions were aimed at evaluating a range of system inputs, outputs and transformation processes also relating to whether inspections:

1. Should be tougher, and more frequent?
2. Should be extended to all flags and ship operations?
3. Result in undue delays, and if so, was there was a better way to tackle the problem of substandard ships?

System Performance: Which Groups were Not Doing their Job Properly?

The systems approach adopted highlighted alternative viewpoints and although indicating little consensus even within one group, revealed an urgent need for education to raise the mutual understanding of all parties. Views were canvassed regarding the performance of each group, and respondents then assigned the proportion of organisations in each group that they felt to be failing to do their job properly as 0, 25, 50, 75 or 100 per cent. It is possible that the reported mean percentages based on these loose bands may exceed the precise values which respondents might have chosen (Table 9.1).

As might be expected, port authorities preoccupied with safety legislation judged most groups more severely than owners, and although no consensus emerged even within one sector, representing widely differing opinions, flag States were cited as the main culprits with all respondents naming at least one negligent State. States that lacked the manpower required to conduct surveys, or possessed insufficient organisational infrastructure to implement legislation, were viewed as avoiding their responsibilities. One respondent believed that open registers encouraged Flags of Convenience states to 'take the money and run'.

Table 9.1 Maritime organisations felt to be 'failing to do their job properly'

Percentage of maritime organisations 'failing to do their job properly' as viewed by:	Ports	SM	Overall
Shipowners and managers, SM	38	31	34
Classification societies, CS	31	25	31
Flag States, FS	51	41	48
Insurers and P&I Clubs, IP	25	31	27
Maritime organisations, MO	27	31	28

Source: Elmdaghri (1994)

In criticising shipowners, one port felt that they invariably chose the cheapest option in manning, maintenance and repair, and a spokesman for the International Transport Workers Federation, argued that when money is tight, owners cut back on safety and maintenance and reduce manning to a minimum. One port felt that 'badly trained, poorly educated but cheap crews are used to replace experienced UK seafarers. Officers from some flag States are poorly qualified and in combination with poor ships, standards are low'. Some shipowners (31 per cent) felt that their colleagues were operating poor quality vessels, creating unfair competition, with one owner stating that other owners were cutting corners for economic gain, and a ship management company felt that some relied on class, with little sense of any self discipline.

Classification Societies were less criticised, particularly by owners, but one manager felt that if more societies joined the International Association of Classification Societies (IACS), fewer non-members would be available to less conscientious owners (e.g. LSM, 1994). One identified an international issue whereby ships that lose their classification could approach another society and gain full rights to sail. Three of the four insurers and P&I Clubs who replied, reliant on classification societies' certificates before providing an insurance policy, felt that societies were failing, with one blaming losses on societies which persisted in issuing certificates in particular classes, allowing defective vessels to trade. Although more tonnage is now concentrated in reputable societies, the need for a culture of co-operation rather than one of blame between IACS and PSC is still apparent (IACS, 1996, 1998), where human rather than technical failings predominate in the sequence of events preceding many casualties. Education and training are integral to establishing this culture, although the willingness by practitioners to upgrade their skills (Dinwoodie, 1999b) does not guarantee that they will do so.

Although less harshly so, some insurers and P&I Clubs were criticised, with one port noting that 'underwriters appear to be motivated by crude capitalism where any vessel, however dilapidated, will eventually be insured, albeit at high premium'. Owners, noting that insurance costs money, observed that bad and good operators obtained similar premiums, with concerns that some insurers were not inspecting ships or liasing with classification societies. A managing director stated that 'insurers are not understanding that their risk is not rewarding good operators', implying a need for greater differentiation of premiums. Of maritime organisations, IMO was criticised for 'lacking teeth', its slow bureaucracy, and has been criticised by Brussels (Moloney, 1994), although several ports and owners were satisfied with maritime organisations.

Attitudes to Substandard Ships

Owners considered that the proportion of substandard tankers and bulkers were similar (30 per cent), as did port authorities (36 per cent). However the limited sample of surveyors and insurers estimated about 50 per cent more bulkers than tankers to be defective, and in 1993, the most recent statistics in the minds of respondents when surveyed, both suffered similar loss ratios. In addition to the 7.98 per cent of bulkers inspected in Paris MOU ports and detained for serious defects and 7.32 per cent of tankers,

52 per cent of bulkers and 43 per cent of tankers revealed at least one defect when inspected. Insurers' concerns regarding sub-standard bulkers may reflect high claims paid out. More recently, the problem of substandard vessels persists, with the UK Maritime and Coastguard Agency still announcing detention rates in excess of nine per cent of the 25 per cent of vessels inspected (MCA, 1998). The gulf between perceived defect levels (31 per cent of vessels) and reality, indicate a need for education to raise awareness of the actual inspection and detention rates.

All but six per cent of respondents agreed or strongly agreed with the statement that there was a problem regarding the number of substandard ships trading. One surveyor highlighted a RoRo ship that had sailed for five days with defective hatch lids. Many differences in perceptions were apparent but a harbour master's view that even one substandard ship was too many is noteworthy.

When asked whether they felt that authorities should draft more regulations, 74 per cent overall felt that they should not do so, rising to 94 per cent for owners who preferred self-regulation, although 15 per cent of port authorities favoured more regulation. Some 98 per cent of all respondents wanted authorities to concentrate on enforcing legislation, including all owners except one, although irresponsible owners were unlikely to have responded to the survey.

The Perceived Role of PSC

PSC was felt to have an important role to play in raising shipping standards by 92 per cent of respondents, but 17 per cent of owners believed that PSC had only a moderate role to play, possibly reflecting their imaginative ability to avoid safety nets. One P&I Club felt that although PSC was a less than ideal vehicle for the eradication of substandard ships and owners, it was surely better than nothing?

In the European Region 43 per cent of respondents felt that the success of the Paris MOU PSC had been significant, and 47 per cent moderately so. Although recently, its effectiveness had grown with increased and higher quality inspections, and better targeting and detention procedures, ships were still escaping due to lack of co-operation, harmonisation and co-ordination.

Of the shipowners surveyed, 56 per cent felt that the Paris MOU had been successful, 33 per cent noted moderate success and 11 per cent no success, possibly reflecting a lack of discrimination between bad and good operators, with a need for better targeting. Some ship managers felt that that surveyors were overworked and preferred to inspect well maintained ships, which took less inspection time, caused less trouble and produced better statistics implying a need to target ships for inspection (Grey, 1994).

Only 30 per cent of port authorities surveyed considered that the Paris MOU had been a success. In practice, deficient ships, once outside territorial waters, could avoid sanctions. Fear that inspections were superficial, or that detention orders were being placed on ships in port for planned repairs were raised. Reticence to detain ships in the face of compensation in cases of undue detention and a need for more funding and greater consistency in training were also expressed. More recently, training for PSC inspectors has become more widespread (IIJNET, 1998b), and UK inspectors now carry identity cards (DETR, 1998b), although there is still room for improvement.

If and How should PSC Powers be Enhanced?

Half of all respondents agreed with the view that PSC should have increased powers, but the responses of both the 70 per cent of ports who favoured increased powers, and 39 per cent of owners who disagreed, fearing additional harassment by inspectors, were predictable. Extension of PSC to other regions of the world was supported by 69 per cent of respondents, with suggestions of a need for an international database that would follow substandard vessels. The establishment of regional PSCs, coupled with inter-regional co-operation, offering global coverage and denying substandard ships the ability to hide by switching trade routes has occurred, with regional databases and inter-regional exchange servers now established. However in this survey, shipowners, perhaps fearing delays in port, were relatively less supportive of the extension of PSC. Generally, eight per cent of respondents disagreed with extended coverage of PSC, possibly due to the inadequate maritime infrastructure or the necessary financial and human resources in other areas. Trained and experienced surveyors may not be available to carry out efficient and effective control of foreign vessels, and some feared a lack of harmonisation between regions, as experienced by the Paris MOU. One view queried how countries that could not properly undertake flag State responsibilities, could busy themselves with the establishment of an efficient Port State inspection regime. There were also fears that PSC in some areas of the world, particularly in developing countries, could facilitate corruption. IMO has facilitated regional co-operation, e.g. resolution A682(17), but some concerns remain. Fears that PSC may be used as a trade barrier or a retaliatory tool were shared by 49 per cent of respondents. Other misgivings indicate the need for more publicity and education regarding the benefits of PSC.

How could Inspection Procedures be Improved?

The 46 per cent of respondents in favour of tougher inspections included one port authority that felt that existing regulations should be applied more forcibly. However, while stricter PSC measures might encourage improved enforcement, owners might equally question the qualifications and experience of the State inspectors. This survey revealed that 77 per cent of owners did not want tougher inspections, but remembering that conscientious shipowners have less to fear from tough inspection, and may positively gain as substandard competitors are compelled to invest in safety, suggests that any additional regulation must be carefully considered. Should inspections be more frequent? Sixty three per cent overall supported this view, which one respondent felt should 'increase checking of certification of crew as well as vessels'. The 59 per cent of owners who did not favour more frequent inspections feared more frequent disruptions by PSC inspectors, but again, responsible owners could benefit as substandard ships are weeded out.

Should inspections cover all flags? All owners, and all but six per cent of respondents, agreed so, thus avoiding discrimination, but targeting could discriminate more between flags, to raise the effectiveness of PSC, and even owners, types of ships, and specific ships, namely those with poor records.

Should inspections cover ship operations? Half of respondents supported this view, with stronger support from ports and maritime organisations, recognising that 80 per cent of marine accidents are attributable to human error. Careful attention to the human factor, checking that crew and officers are qualified, competent, and able to carry out their responsibilities with a minimum of proficiency, and that officers and crews were able to communicate amongst themselves could reduce accident rates. On the other hand, the remaining half did not support control being extended to operations, particularly true of shipowners, with 71 per cent against these extended controls. It has been alleged that it would be very difficult to assess competency of crews in a short time period, especially if inspectors did not share a common language.

Furthermore, it was argued that this area of control would result in subjective judgements being made.

Will inspections result in undue delays? Threats of compensation claims against surveyors if vessels are excessively detained or delayed, might help to explain why 54 per cent of respondents considered this not to be so. Alternatively, those holding this view might sec control as a necessary evil, resulting only in necessary delays to shipping movements. A minority (27 per cent) of respondents who did consider that inspections could result in undue delays, still believed that PSC was needed, since all respondents who thought undue delays may occur also considered that PSC was important.

Are there Better Ways to Tackle the Problem of Substandard Ships?

Many respondents (52 per cent) believed that there were better solutions, but 42 per cent of ports did not. Port authorities noted a need for greater willingness to detain vessels and prosecute offenders, and better education. The majority (72 per cent) of owners considered that better means than PSC were available to deal with the problem of substandard ships, but half of these were satisfied with the memorandum.

For a substandard ship to exist there is a chain of events, and according to one managing director 'PSC can only be part of the solution to tackle the problem', and each link in the chain could play its part in raising standards. One surveyor believed that greater intervention by underwriters would assist. One organisation supported 'commercial incentives for reliable operators'. One harbour authority favoured better flag State control, and removal of registries of convenience. According to one P&I Club: 'The sovereign right to create a flag State will continue to be a critical weakness in the system, unless the international community, through IMO, can create and enforce standards for flag State authorities.' The majority who had better solutions in mind still supported PSC and considered its role as relevant.

Conclusions

A soft systems approach was used to study the varying views of shipowners, classification societies, insurers, and flag States who by failing to discharge their responsibilities properly, may be lowering safety standards at sea, exposing PSC as the remaining safety net (Plaza, 1994).

Despite numerous rules and regulations, it was found that substandard vessels have created an unfair competitive environment where conscientious shipowners are struggling to survive.

Perceptions of the extent of the problem vary. In 1992, of 6000 foreign vessels visiting the UK, 60 per cent of the 2000 inspections conducted revealed deficiencies, with six per cent resulting in detaining the ship. These published detention rates, which influenced the attitudes of respondents as surveyed in 1994, exceed nine per cent in 1998, and when combined with high levels of shipping accidents and fatalities, have tarnished the image of the shipping industry. PSC, created in Europe by the Paris MOU, a response to tackle substandard shipping, became necessary when controls conducted by some flag States were unsatisfactory. The Paris MOU facilitated detection and detention of substandard ships, but global co-operation, co-ordination and harmonisation is still needed to prevent rogue ships from hiding on less supervised routes. An educational programme, applicable to a broad range of interested practitioners presents one means of raising the awareness of the incidence and frequency of both deficiencies and detentions, and the causes and consequences of accident losses.

How could the need for PSC be reduced? Despite 15 years since the Paris MOU, the proportion of substandard vessels has changed little. Many concerns expressed by respondents in 1994, which supported Lord Donaldson's (1994) findings that PSC was a weak deterrent, with surveyors limited by an inability to deal with passing traffic, merely ensuring the validity of ships' certificates and inability to discover latent deficiencies, have been addressed. However, to varying degrees, problems still remain relating to the threat of a claim for compensation, limited resources both financial and human, lack of follow-up action, lack of harmonisation, varying standards of individual inspectors, varying severity and efficiency of PSC from country to country, and lack of discrimination. Inadequate training or resources for port development remain in many developing countries (Thomas, 1995; Mobarek, 1997), where PSC systems must compete for resources amongst other priorities. Targeting and detention procedures have improved, but ineffective penalties may still hamper the effectiveness of PSC, with some shipowners apparently still concluding that the balance of commercial advantage lies in ignoring the effects of the Paris MOU. This UK based survey reported here confirmed that PSC has a significant role to play in raising shipping standards. While respondents were divided about the need for tougher inspections, and inspections including ship operations, the majority was in favour of more frequent inspections covering all flags. However, it is also apparent that academic educational programmes in marine studies must prepare students, via role plays, simulations and group exercises, to grasp the importance of taking responsibility personally for all their future professional actions. The long term solution to reducing both substandard ships and consequent accidents surely depends as much on the attitudes and willingness of practitioners to undergo continuing education and training as well as on the technical competence of all the professional mariners involved.

Could operational changes raise the effectiveness of PSC inspectors? Since this survey was conducted, recommendations and suggestions aimed at strengthening the effectiveness of PSC have been implemented in varying degrees, with progress on the need for increased harmonisation, co-ordination and closer co-operation both within and between regions, but more resources still need to be earmarked for PSC. A levy system, involving a nominal inspection fee and fines proportionate to the magnitude of any deficiency found could be developed. Greater disincentives and sanctions (HCB, 1993) might include detention coupled with scaled fines based on the seriousness of deficiencies found with a denial of the right to load and unload for persistent offenders, or conduct-related insurance premiums. Better targeting of vessels with poor detention records is being implemented, along with more elaborate detention procedures and better training and clear guidelines for inspectors. Inspections are possible where 'clear grounds' exist for detaining ships, lists of detained ships are publicised (Donaldson, 1994) naming the parties involved, and greater sharing and disclosure of information between agencies is occurring. However, some flag States still need encouragement in ratifying the IMO Conventions that they have not yet done so, and to invest more resources into improving their maritime infrastructure. The soft systems approach adopted here suggested that enhanced training of inspectors, although potentially useful, was likely to be of lower priority than a need to ensure more responsible actions by other groups involved. Devising an appropriate educational strategy to assist them is a greater priority.

The next phase of improvements will demand the development of a safety culture both on ships and amongst those managing them, predicated by improved education of all parties. This will focus on raising awareness of the:

1. Incidence, frequency, causes and consequences of accidents and substandard ships.

2. Importance of a responsible approach by each agency and an appreciation of the role of other agencies in ensuring safe operations.
3. Development of a commitment to safe operations.
4. Role of safe operations in ensuring a favourable public response to shipping issues.
5. Need for industry wide co-operation and agreements promoting safe operation.

Such education, aimed at honing the 'human element' in shipping, must extend far beyond intense short training courses for inspectors. Rather, it will involve shipowners and port authorities, insurers and inspectors, working together on joint exercises in the classroom, throughout their careers. These exercises must encourage self-discipline in their participants, and be designed to remove suspicion between functional disciplines within the business of shipping, until a genuine safety culture becomes endemic, worldwide.

10 Implications for Practitioners in Shipping and Logistics

Why are the Views of Students Important?

Today's advanced students will be tomorrow's managers. As such, the research reported above provides a unique insight into the interests and concerns of the shipping and logistics managers of the future, in terms of their current career and educational aspirations. In posing the vital question of why postgraduate and other advanced students of shipping and logistics chose to embark on their courses in Britain, they reveal views not only about themselves, but also about the current state of their industries and educational provision within it. Some of the implications of findings for a range of practitioners, both industrial and academic, may be unpalatable, but they may also hold the key to a more solid future.

An important lesson to learn is that each individual is different, and demands an educational system that treats them as such, with short shrift given to institutions perceived as failing to do so. No single overriding issue predominated for all students, but rather, a complex web of considerations operated even for one individual when contemplating further study. Depending on the academic discipline involved, be it shipping or logistics, and whether the individual already possessed prior industrial experience or otherwise, their particular motivations for undertaking study varied substantially.

This final chapter addresses the aim of suggesting methods by which the management of vocational marketing and recruitment campaigns might be improved, and recommending ways to promote recruitment into advanced vocational courses whilst maintaining high entry standards. Many of these issues have already been alluded to, but the remaining comments are directed at particular groups of practitioners and individuals in the hope that they will be enticed to reconsider what has gone earlier.

Some Implications for...

...Employers and Operators in Shipping and Logistics...

There is no such entity as 'the shipping industry', but rather a range of specialist activities and occupations which involve interaction with the sea to varying degrees. In the present context, shipping companies, ship management companies, port authorities, regulatory bodies, consultants, professional bodies and many more could be included. All need to known that the features common to most of the students of shipping studied here, were a 'love of the sea' which presented a major appeal of marine related courses, and the perceived vocational credibility of their qualifications. Positive actions by any of these groups to reinforce these perceptions are likely to encourage the recruitment process and hence assist in finding and developing their future leaders. Sponsorship of students on courses by employers, provision of sandwich placements, short term training or 'taster' places in the workplace, a willingness to make presentations in schools, universities and at careers events cost little to provide. However, they can have a major effect on the formative perceptions of receptive young people and more subtly, public perceptions of both the organisation and the industry it represents. The long term effects of recruitment videos, seminars for careers advisers, school packs, business games and specialist careers guides will need careful monitoring, but are surely a small overhead to pay to secure a vibrant professional recruitment regime.

At the level of assessing corporate training provision for employees, this work has shown that specialist techniques exist which can assist managers in the process of structuring complex decision situations and thinking through their training priorities. A worked example of applying the technique of cognitive mapping (Chapter 6) to a decision relating to where to fund staff to attend short courses should be accessible to most managers in many similar situations. At a much deeper level, indicative values of indices comparing the views of different individuals but based on the same technique and study decision were presented. These showed clear differences in perceptions of the study decision by shipping and logistics students, but gender had little effect. Such techniques are probably most useful for comparing the cognitive maps of just a few individuals or groups rather than the large numbers shown here. This in itself illustrates a simple lesson that if human resource development resource allocation decisions can respond to clear statements of perceived needs by the individuals concerned, based on their own words or maps (Chapters 4 and 6), it is more likely to be effective in meeting those needs.

...A Practitioner Considering a Return to Study...

Chapter 1 reported on declining ship officer manning levels, with trends towards larger and faster ships placing ever greater financial and operating responsibilities on all personnel including seagoing officers and staff in shipping companies, shore based operations and supporting services. Ever more able and qualified managers are required to run these activities. Coupled with this, poor and sensitive public images in both shipping and logistics have influenced the ability to recruit high quality staff in some advanced economies. But however strong the forces of supply chain re-engineering are, unmanned ships will not be commonplace short term, and strategic realignment of supply chains will not result in extensive air, road or rail movements or relocation of sources supplanting sea transport systems.

If you are a mid career practitioner seeking to go ashore, or currently in industry seeking to upgrade or update knowledge, and considering gaining post-experience or postgraduate qualifications, then do it. There are many openings for experienced mariners in organisations ashore where seagoing experience, tempered by recent academic attainment, is valued like gold dust. Equally, the modern classroom can offer vital 'time-out' to gain new academic knowledge and understanding, acquire enhanced interpersonal skills often in an international environment, and make new personal contacts in an informal atmosphere. Appendix 1 reports on the issues raised by practitioners who had recently returned to study, and Chapter 5 openly reports the extent of perceptions of the attractions and deterrents to study. A realistic prior assessment of any potential barriers or possible downside to study as reported here, and also of possible future occupations not reported here, can strengthen resolve and raise the chances of achieving real satisfaction and success. And it really is worth doing lots of research into what courses are available, because as one student noted, he couldn't afford to repeat the experience.

...A Person Interested in Going to University or a Career at Sea...

Professional training for seagoing officers, on deck or in the engine room, has never been so geared to preparing participants for not only their days at sea, but also any future career beyond. Academic business and law studies incorporated into many courses for training deck officers now provide not only a working knowledge of immediate vocational practices, but also a respected academic grounding. Modern courses for ships' engineers also provide technical knowledge and skills applicable far beyond the shoreline. The personal qualities and interpersonal skills that any seagoing career relies on are highly valued by businesses worldwide, and former seafarers are rarely left high and dry at the quayside.

So what of university courses? A course in marine studies may well be located near the sea, which is convenient if you are seeking a pleasant environment in which to study, or are interested in watersports. You will probably have an academic interest in the sea. You may know friends or relatives who have been at sea, or might be attracted by the lure of travel, dynamism and the buzz of a fast-changing business world. Experience suggests that you will become much more concerned with the basic realities of employment, salaries and prospects nearer graduation. It is certainly helpful to find out as much about relevant occupations and gain 'taster' experience in those that look attractive as early as possible. But again, be realistic. Eventually, the attraction of spending a couple of days at the other side of the world instead of being home at weekend might wane. If it does, has your course prepared you for a change in lifestyle?

Does an international career appeal to you? If so, consider a course offering studies in an international classroom, with the possibility of time spent abroad. The personal contacts and skills gained in the classroom will be apparent to any future employer, but only you will be aware of the self-confidence the experience has given you. Study in the UK is a unique attraction to many aspiring managers in maritime industries worldwide, in the land of the language and traditions of shipping. Chapter 2 discusses some of the issues involved in really preparing for international careers.

...Human Resources Managers...

For the qualified practitioner seeking to update or merely maintain a continuing professional development record demanded by a professional body, appeal to issues of ongoing professionalism, pride and excellence in practice can generate genuine satisfaction. The soft systems methodology presented in Chapter 9 should be accessible to human resources managers and potential providers of training services in a range of occupations, as a means of tailoring personal development programmes to meet specific industrial needs. However, these are two-edged swords. The approach towards devising a training strategy for inspectors of port state control could be modified to suit many differing occupational groups in shipping and logistics, with each possessing its own requirements. Conversely, issues of awareness, understanding and co-operation between agencies demand much broader professional involvement in development programmes. Merely training inspectors does little to hone the attitudes and culture of owners and operators. Full-time educational courses could usefully address these issues explicitly providing a career-entry foundation which individuals upgrade through regular participation in multi-disciplinary professional conferences, workshops and courses.

In a fast moving international environment, continuing professional development is essential for all managers. Research centred on Plymouth for example revealed postgraduate education in shipping and logistics to be a genuinely international business, with a typical class comprising participants from all continents. In general, less than ten per cent of students surveyed in this group were British, with a majority of European origin, but including varying proportions from Northern and Southern Europe. About 20 per cent of students were not of European origin. The multinational classroom presents an unparalleled forum within which to develop personal and social skills, form new relationships and even effect the behavioural changes, essential for effective management interaction in an international environment. It is vital for promoting corporate survival and growth that as many practitioners as possible return as frequently as possible to share in these exchanges, and that such a unique experience is regarded as a recognised prelude to an international career.

The value of ongoing vocational guidance is self-evident where large proportions of the postgraduate students surveyed were observed to possess experience of the workplace, and also where basic employment motives were found to be important to almost all students. Such guidance might address student concerns relating to potential exit strategies, both for applicants moving from the workplace into the classroom, and vice versa for graduates, helping individuals to form realistic expectations of their future prospects. Additionally, the reflective processes associated with lifelong learning need to be fostered within in a critical environment of growing self-awareness.

A majority of the postgraduates surveyed in this work were concerned that potential employers would view their awards as unique, and that they would only seek to apply to study at academic institutions with established and recognised reputations. This concern for 'credentialism' hints at a means of appealing to aspiring managers to upgrade their professional qualifications, within the context of raising industrial standards generally.

In the process, appeals to pride and status, challenge and vocational development could usefully be viewed within a framework of a holistic approach to personal development.

The perceived importance of prior academic links, and the need to maintain them after graduation, was evident in this work. Not only does this afford scope for developing a genuine culture of lifelong learning, but also accentuates the role of alumni in the workplace as potential recruiting sergeants, which may have been underestimated traditionally. The desire to study, and to continuously upgrade must become endemic, as part of a pro-active culture in which the individual is encouraged and enabled to realise their full personal potential.

...Providers and Marketers of Courses...

Substantial but varying proportions, nationally, of matriculants to the vocational courses investigated in this work claimed to possess relevant prior work experience. This situation implies a need to:

1. Recognise the true value of professional experience in postgraduate course entry requirements.
2. Devise marketing materials which entice mature individuals through emphasising the vocational continuity of their prior work experience and their pending return to the classroom.
3. Encourage course members to reflect on and share their practical knowledge during the pedagogic process.

In terms of how prospective students found out about courses, 'talking to people in industry' was found to be less important to postgraduates than undergraduates, as was 'talking to lecturers'. Both findings point to the importance of practitioners and lecturers being open to approaches from schools and colleges and of appearing to be genuinely enthusiastic when approached. However, postgraduates rely on other methods for finding out about courses. In that it influenced almost 40 per cent of both undergraduates and postgraduates, the role of 'family and friends' was an important one. Whilst this suggests a conservative industry, often influenced by family traditions, an obvious marketing implication is to raise the general profile of the attractions of careers in shipping and logistics to attract more potential applicants. However, only a small proportion of those surveyed here claimed this to have been a major influence on them, and the importance of helping each individual to feel personally in control of their own decision to study is particularly important at postgraduate level.

As reported in Chapter 3 a schema of perceptions was found to guide student conceptions of vocational employment roles. Vocational awareness at matriculation is often low, and lecturers need to experiment with methods such as those described that may succeed in accelerating this awareness. As students become more vocationally aware, such research could then usefully turn to finding ways of raising student awareness of the detailed skills and knowledge demanded in a chosen subset of roles.

...Academic Researchers...

The research methods employed commenced with qualitative analysis to develop a research instrument. Following testing, this was applied over three different cohorts at one university, and with slight modifications, nationally. In an under-researched area, using qualitative data collection methods enabled international students to express their personal concerns in their own words. Similar approaches are likely to be needed to establish the concerns of particular groups of respondents.

How far can findings be extended to other universities, and in related academic disciplines such as international business? Further research is needed, but some tentative conclusions are useful. Firstly, there was little evidence of major changes in the reminiscences of students as tested at three and six months after matriculation, in a control group, with perceptions of most issues being observed to remain relatively stable. However the periods of elapsed time between a decision to apply having being made, a final commitment taking place and arrival at university varied between individuals. Further study of the very detailed evolutionary processes whereby decisions are made, and how perceptions are influenced by more immediate events during the decision period are desirable. However, given that it is not possible to record in detail the situations of large numbers of prospective graduates, or potential mid career returners to education, detailed ethnographic or other monitoring of individuals before they decide to make applications to study is infeasible. Small scale study of a captive audience of prospective undergraduates whilst still at school has proved possible and fruitful (Moogan et al, 1999), and may be feasible amongst final year undergraduates. However if few of those surveyed actually proceed into particular postgraduate courses this would imply a substantial wasted effort. In the short term, research is probably best concentrated on using reminiscence data from a range of institutions, gathered as soon after matriculation as possible.

Do related subject areas attract similar concerns? It remains unknown how far students of international business, general transport or related areas of technical logistics or marine studies exhibit similar motivations for study. Just as the cognitive maps of shipping and logistics students varied, so further distinctions between students on academic and vocational courses might be expected. Only additional research can define the extent of influences such as the status afforded to UK qualifications, or the role of family and friends in the study decision.

Finally, in other cultures, other influences on students may be at work. In particular, the regard given to individual aspirations may be overtaken by cultural or economic considerations. In such environments the importance of the cognitive maps of individual students may be subsumed by those of their sponsors, be they corporate, family or state, who fund studies. Further research in other countries is needed to reveal this possibility.

Bibliography

Ackermann, F.R., Cropper, S.A. and Eden, C.L. (1991), 'Cognitive mapping for community operational research - a user's guide', in A.G. Mumford and T.C. Bailey (eds.), *Operational Research: Tutorial Papers, 1991*, Operational Research Society, Birmingham, UK, pp. 37-52.

Adler, N.J. and Ghadar, F. (1990), 'Strategic human resource management: a global perspective', in R. Peiperl (ed.), *Human Resource Management: An International Comparison*, Walter de Gruyter, New York, pp. 236-60.

Ainsworth, M. and Morley, C. (1995), 'The Value of Management Education: Views of Graduates on the Benefits of doing an MBA', *Higher Education*, vol. 30, pp. 175-187.

Baldwin, T.T., Magjuka, R.J. and Loher, B.T. (1991), 'The perils of participation: effects of choice of training on trainee motivation and learning', *Personnel Psychology*, vol. 44, pp. 51-65.

Bartlett, C. and Ghoshal, S. (1992), 'What is a global manager?', *Harvard Business Review*, September-October.

Board of Trade (1970), *The Attitude of Seafarers to their Employment*, The Committee of Inquiry into Shipping, Social Surveys (Gallup Poll) Ltd., Board of Trade, London, UK.

Bogoun, M.G. (1983), 'Uncovering cognitive maps. The Self-Q Technique', in G. Morgan (ed.), *Beyond Method*, SAGE Publications, Beverly Hills, California, pp. 173-188.

Boisson, P. (1994), 'Classification societies and safety at sea: back to basics to prepare for the future', *Marine Policy*, vol. 18, no. 5, pp. 363-377.

Boreham. C. and Arthur, A.A. (1993), *Information Requirements in Occupational Decision Making*, Employment Department Group, School of Education, University of Manchester, UK, Research Series No. 8.

Boulter, J.D. (1991), *Recruiting: Marketing the Merchant Navy*, unpublished DMS project, Plymouth, UK, Institute of Marine Studies, University of Plymouth.

Boyer, E.L. (1990), *Scholarship Reconsidered: Priorities of the Professoriate*, Carnegie Foundation for the Advancement of Teaching, New Jersey.

Breakwell, G.M., Hammond, S. and Fife-Schaw, C. (1995), *Research Methods in Psychology*, SAGE Publications Ltd., London, UK.

Brennan, J., Kogan, M., and Teichler, U. (1996), 'Higher education and work', *Higher Education Policy Series*, no. 23, Kingsley, London, UK.

Bumstead, J. (1998), 'Time Compression in the supply chain: compress your supply chain: expand your customers' satisfaction', in J. Gattorna (ed.), *Strategic Supply Chain Alignment, Best Practice in Supply Chain Management*, Gower, Aldershot, UK, pp. 157-170.

Burn, B.B., Cerych, L. and Smith, A. (1990), 'Study abroad programmes', *Higher Education Policy Series*, no. 11, vol. 1, Kingsley, London, UK.

Clinton-Davis, Lord (1994), 'The UK, Europe and the global maritime trading system', *Marine Policy*, vol. 18, no. 6, pp. 472-475.

Conrad, L. and Phillips E.M. (1995), 'From isolation to collaboration: A positive change for postgraduate women?', *Higher Education*, vol. 30, pp. 313-322.

Counsell, D. (1996), 'Graduate careers in the UK: an examination of undergraduates' perceptions', *Career Development International*, vol. 1, issue 7, pp. 45-51.

Cownie, F. and Addison, W. (1996), 'International students and language support: a New Survey', *Studies in Higher Education*, vol. 21, pp. 221-231.

Crabtree, B.F., Yaboshik, M.K., Miller, W.L. and O'Connor, P.J. (1993), 'Selecting individual or group interviews', Chapter 9 in D.L. Morgan (ed.), *Successful focus groups. Advancing the State of the Art*, Sage Publications, London, UK.

Cramer, D. (1997), *Fundamental Statistics for Social Research: Step-by-step Calculations and Computer Techniques using SPSS for Windows*, Routledge, London.

Daley, J.M., Murphy, P.R. and Smith J.E. (1994), 'Ethics education and the transportation industry: Transferring classroom techniques into management education', in *Proceedings of the Intermodal Distribution Education Academy*, pp. 7-15.

Dearing, R. (1997), 'Full and part time students in higher education: their experiences and expectations', Report 2, *Higher Education in the Learning Society*, National Committee of Education into Higher Education, HMSO, London, UK.

DETR (1997), Department of Environment Transport and the Regions, *Transport Statistics Great Britain 1997 Edition*, The Stationery Office, London.

DETR (1998a), Department of Environment Transport and the Regions, *Transport Statistics Great Britain 1998 Edition*, The Stationery Office, London.

DETR (1998b), *Statutory Instrument 1998 No. 1433*, The Merchant Shipping (Port State Control) (Amendment) Regulations 1998, regulation 7, UK Department of the Environment Transport and the Regions, 9th June.

Dijxhoorn, O.H.J. (1993), 'Port and shipping management: the role of IMO', *Marine Policy*, vol. 17, no. 5, pp. 363-366.

Dinwoodie, J. (1994), *Transport as a Career: A Case-Study at the University of Plymouth*, unpublished Master of Education (Higher Education) thesis, Faculty of Arts and Education, Exmouth, University of Plymouth, UK.

Dinwoodie, J. (1996), 'The decision to study transport at university', *Proceedings of the Chartered Institute of Transport*, vol. 5, no. 2, pp. 46-54.

Dinwoodie, J. (1999a), 'Learning through accident or academy? The potential of classroom based study for developing competencies in managers of ferry services', in F. Yercan (ed.), *Ferry Services in Europe*, Ashgate, Aldershot, England, pp. 118-149.

Dinwoodie, J. (1999b), 'Safety, attitudes of practitioners to updating courses and the role of continuing maritime education in an era of lifelong learning', in International Maritime Lecturers Association, *Conference on Maritime Education and Training: Preventing Accidents, Dealing with Emergencies, Coping with Casualties - The Education and Training Perspective*, Opatija, Croatia, May 1999, vol. 1, pp. 93-103, Rijeka College of Maritime Studies, Rijeka, Croatia.

Dinwoodie, J. (1999c), 'The pedagogic value of a computer based instrument investigating why students embarked on postgraduate business courses', in K. Fletcher and A.H.S. Nicholson (eds.), *Selected Proceedings from the 10th Annual CTI-AFM Conference, Brighton, UK, April 1999, CTI-Accounting, Finance & Management*, University of East Anglia, Norwich, UK, pp. 181-189.

Dinwoodie, J. (2000), 'The perceived importance of employment considerations in the decisions of students to enrol on undergraduate courses in Maritime Business in Britain', *Maritime Policy and Management*, vol. 27, no. 1, pp. 17-30.

Dinwoodie, J. and Heijveld, H. (1997), 'Ensuring quality managers in the marine industry: an analysis of the undergraduate decision to embark on a marine studies course', in O.K. Sag (ed.), *Proceedings of the Eighth Congress of the International Maritime Association of the Mediterranean, Volume 3, Istanbul, Turkey*, November 1997: Istanbul Technical University Maritime Faculty, pp. 13.1.1-13.1.10.

Donaldson, Lord (1994), *Safer Ships, Cleaner Seas*, report of Lord Donaldson's inquiry into the prevention of pollution from merchant shipping, HMSO, esp. pp. 135-155 and 374-378.

DTp (1984), *Transport Statistics Great Britain (1973-1983)*, Department of Transport, HMSO, London.

DTp (1988) *Transport Statistics Great Britain (1988 Edition)*, Department of Transport HMSO, London.

DTp (1993), *Transport Statistics Great Britain (1993 Edition)*, Department of Transport HMSO, London.

Dunn, W.N., Cahill, A.G., Dukes, M.J. and Ginsberg, A. (1986), 'The policy grid: a cognitive methodology for assessing policy dynamics', in W.N. Dunn (ed.), *Policy Analysis: Perspectives, Concepts and Methods*, pp. 355-375, JAI press, Greenwich, CT.

Eden, C. (1990), 'Strategic thinking with computers', *Long Range Planning*, vol. 23, pp. 35-43.

Eden, C. and Ackermann, F. (1993), 'Evaluating strategy - its role within the context of strategic control', *Journal of the Operational Research Society*, vol. 44, pp. 853-865.

Elmdaghri, H. (1994), *Attitudes of the UK Shipping Industry Towards Port State Control and its Role in Raising Safety Standards at Sea*, unpublished Masters in International Shipping, Institute of Marine Studies, University of Plymouth thesis, Plymouth, UK.

Entwistle, N., Thompson, S. and Tait, H. (1992), *Guidelines for Promoting Effective Learning in Higher Education*, Centre for Research on Learning and Instruction, University of Edinburgh.

Evangelista, P. and Morvillo, A. (1998), 'The role of training in developing entrepreneurship: the case of shipping in Italy', *Maritime Policy and Management*, vol. 25, no. 1, pp. 81-96, esp. p.93.

Evans, J.T. (1989), *Bias in Human Reasoning: Causes and Consequences (Essays in Cognitive Psychology)*, Lawrence Erlbaum Associates Ltd., Hove, UK.

Ferch, S. and Roe, M. (1998), *Strategic Management in East European Ports*, Plymouth Studies in Contemporary Shipping, Ashgate, Aldershot, UK.

Fish, A. and Wood, J. (1997), 'Realigning international careers - a more strategic focus', *Career Development International*, vol. 2, issue 2, pp. 99-110.

Frickle, P.H. (1974), *Social Structure of Crews of British Dry cargo Merchant Ships: A Study of the organisation and environment of an occupation*, Cardiff, Wales: Department of Maritime Studies, University of Wales, Institute of Science and Technology.

Fulmer, R.M. and Gibbs, P.A. (1998), 'Lifelong learning at the corporate university', *Career Development International*, vol. 3, issue 5, pp. 177-184.

Goetz, P.W. (1990), *The New Encyclopaedia Britannica Volume 28 Macropaedia, Knowledge in Depth*, Encyclopaedia Britannica Inc., Chicago.

Gold, J. (1998), 'Telling the story of organisational effectiveness', *Career Development International*, vol. 3, no.3, pp. 107-111.

Goldstein, I.L. and Gilliam, P. (1990), 'Training system issues in the year 2000', *American Psychologist*, vol. 45, pp. 134-145.

Grey, M. (1994), 'Worries abound as Port State Control goes global', *Maritime Asia*, Lloyd's List, January, p. 19.

Hall, D.T. (1992), 'Career indecision research: conceptual and methodological problems', *Journal of Vocational Behaviour*, vol. 41, pp. 245-250.

Hammond, N. (1996), 'Concept mapping as directed reflection', in *CiP, Computers in Psychology, Conference Proceedings* on CD Rom, 1996.

Hansen, C.D. and Willcox, M.K. (1997), 'Cultural assumptions in career management: practice implications from Germany', vol. 2, issue 4, pp. 195-202.

Harris, R. (1995), 'Overseas students in the United Kingdom university system', *Higher Education*, vol. 29, pp. 77-92.

HCB (1993), 'The 13 step method', *Hazardous Cargo Bulletin*, February, vol. 14, no. 2, p. 72, Intapress, London.

HESA, (1997), *Students in Higher Education Institutions 1995/96*, Data Report, Higher Education Statistics Agency, Cheltenham, UK.

Hesketh, A.J. and Knight, P.T. (1999), 'Postgraduates' choice of programme: helping universities to market and postgraduates to choose', *Studies in Higher Education*, vol. 24, pp. 151-163.

Hibbs, J. (1988), *An Introduction to Transport Studies*, 2nd edition, Kogan Page, London, UK.

Hope, R. (1980), *The Merchant Navy*, Stanford Maritime, London, UK.

IACS (1996), *IACS Presentation Flags Closer Links with PSC* Press Release, 3rd April, at http://www.iacs.org.uk/pressrel/1996B4/PSCPR.html.

IACS (1998), *IACS Echoes Working Partnership with PSC*, Press Release, 26th March, at http://www.iacs.org.uk/pressrel/1998/PSC.html.

IIJNET (1998a), *Port State Control in Other Regions*, at http://www.iijnet.or.jp/tokyomou/ar-1-7.html.

IIJNET (1998b), *Training and Seminars for PSC officers*, at http://www.iijnet.or.jp/tokyomou/ar-1-6.html.

Institution of Highways and Transportation (1997), *Transport in the Urban Environment*, IHT, London.

Jochems, W., Snippe, J., Smid, H.J. and Verweij, A. (1996), 'The academic progress of foreign students: Study achievement and study behaviour', *Higher Education*, vol. 31, pp. 325-340.

Johnston, S. (1995), 'Building a sense of community in a research master's course', *Studies in Higher Education*, vol. 20, pp. 279-291.

Kelly, G. (1955), *The Psychology of Personal Constructs*, Norton, New York.

Kidd, J.M. and Killeen, J. (1992), 'Are the effects of careers guidance worth having? Changes in practice and outcomes', *Journal of Occupational and Organisational Psychology*, vol. 65, pp. 219-234.

Knapper, C.K. and Cropley, A.J. (1991), *Lifelong Learning and Higher Education*, 2nd edition, Kogan Page, London, UK.

Kogan, M. (1994), 'Assessment and productive research', *Higher Education Quarterly*, vol. 48, pp. 57-67.

Kruskal, W.H and Wallis, W.A. (1952), 'Use of ranks in one-criterion variance analysis', *Journal of the American Statistical Association*, vol. 47, pp. 538-621.

Lane, A. and Kahveci, E. (1999), 'The formation and maintenance of transnational seafarer communities', found on 29th November at the website: http://www.transcomm.ox.ac.uk/www/root/lane.htm, 2 pp.

Langfield-Smith, K. and Wirth, A. (1992), 'Measuring differences between cognitive maps', *Journal of the Operational Research Society*, vol. 43, no. 12, pp. 1135-1150, Pergamon Press, London, UK.

Leavesley, J. (ed.), (1995), *Occupations*, Careers and Occupational Information Centre, Bristol.

Lent, R.W., Brown, S.D. and Hackett, G. (1994), 'Towards a unifying social cognitive theory of career and academic interest, choice and performance', *Journal of Vocational Behaviour*, vol. 45, pp. 79-122.

Levine, D.M., Berenson, M.L. and Stephan, D. (1997), *Statistics for Managers using Microsoft Excel*, Prentice Hall, New Jersey.

Li, K.X and Wonham, J. (1999), 'Who mans the world fleet? A follow-up to the BIMCO / ISF manpower survey', *Maritime Policy and Management*, vol. 26, no. 3, pp. 295-303.

Linehan, N. (2000), *Senior Female International Managers: Why So Few?* Ashgate, Aldershot, England.

Longenecker. C.O., Simonetti, J.L. and LaHote, D. (1998), 'Increasing the ROI on management education efforts', *Career Development International*, vol. 3, issue 4, pp. 154-160.

LSM, (1994), 'Under attack - Classification societies continue to suffer for their failure to identify substandard ships', *Lloyds Ship Manager*, March, p. 70.

Mann, H.B. and Whitney, D.R. (1947), 'On a test of whether one or two random variables is stochastically larger than the other', *Annals of Mathematical Sciences*, vol. 18, pp. 50-60.

May, A.S. (1997), 'Think globally, act locally! Competences for global management', *Career Development International*, vol. 2, issue 6, pp. 308-309.

MCA (1998), *Detention List*, Maritime and Coastguard Agency, UK, June.

McConville, J. (1999), 'Editorial: Maritime manpower', *Maritime Policy and Management*, vol. 26, no. 3, pp. 207-208.

McConville, J., Glen, D. and Dowden, J. (1998), *United Kingdom Seafarers Analysis, 1997*, The Centre for International Transport Management, London Guildhall University.

McWhirter, E.H. (1997), 'Perceived barriers to education and career: ethnic and gender differences', *Journal of Vocational Behaviour*, vol. 50, pp. 124-140.

Mills, J., Phillip, J. and Murray, W. (1999), 'Attracting More Managers into Logistics', Human Resources Forum in Logistics, University of Huddersfield, UK, 40 pp.

Millward, L.J. (1995), 'Focus Groups', in G.M. Breakwell, S. Hammond and C. Fife-Schaw (eds.), *Research Methods in Psychology*, Chapter 18, Sage Publications, London, pp. 274-292.

Mobarek, I. (1997), 'Ports in developing countries: can they meet the challenge?', in *Proceedings of the 13th International Conference on Port Logistics on 'Port Strategy and Development'*, Port Training Institute, Arab Academy for Science and Technology and Maritime Transport, Alexandria, Egypt, pp. 5.1-5.15, esp. pp. 4-5.

Moloney, S. (1994), 'Brussels raises substandard shipping stakes', *Lloyd's List*, 14th April, p. 5.

Moogan, Y.J., Baron, S and Harris, K. (1999), 'Decision-making behaviour of potential higher education students', *Higher Education Quarterly*, vol. 53, no. 3, July, pp. 211-228.

Moreby, D.H. (1998), *The Economics of Training Ships' Personnel in Britain*, paper first delivered in 1968, Department of Maritime Studies and International Transport, University of Wales at Cardiff, UK.

Moreby, D.H. and Springett, P. (1990), *The UK Shipping Industry - Critical Levels Study*, British Maritime Charitable Foundation, London.

Morgan, D.L. (ed.) (1993), *Successful Focus Groups: Advancing the State of the Art*, Sage Publications, London, UK.

Murphy, P.R. and Daley, J.M. (1997), 'Examining international freight forwarder services: the perspectives of current providers and users', *Journal of Transportation Management*, vol. 9, no. 1, pp. 19-27.

Nicholson, N. and West, M.A. (1988), *Managerial Job Change: Men and Women in Transition*, Cambridge University Press, Cambridge.

Nisbett, R. and Ross, L. (1980), *Human Inference: Strategies and Shortcomings of Social Judgement*, Prentice Hall, Englewood Cliffs, New Jersey.

Obando-Rojas, B., Gardner, B. and Naim, M. (1999), 'A system dynamic analysis of officer manpower in the merchant marine', *Maritime Policy and Management*, vol. 26, no.1, pp. 39-60.

Otala, L. (1994), 'Implementing lifelong learning through industry-university partnership', *Industry and Higher Education*, vol. 8, no.4, pp. 201-207.

Panayides, P. and Dinwoodie, J. (1999), 'The decision to study international shipping at postgraduate level: how far does the evidence support recent socio-cognitive theories of career development?', *First International Congress on Maritime Technological Innovations and Research, Proceedings April 21-23, Barcelona 1999*, pp. 755-776, Department of Nautical Science and Engineering, Universitat Politecnica de Catalunya, Barcelona, Spain.

Parry, G. (1997), 'Patterns of Participation in Higher Education in England: A Statistical Summary and Commentary', *Higher Education Quarterly*, vol. 51, pp. 6-28.

Payoyo, P.B. (1994), 'Implementation of international conventions through port State control: an assessment', *Marine Policy*, vol. 18, no. 5, pp. 379-392.

Piaget, J. (1967), *Biology and Knowledge*, University of Chicago Press, Chicago, p. 18.

Plaza, F. (1994), 'Port State Control: towards global standardisation', *IMO News*, no. l, p. 13.

Pritchard, R.M.O. (1994), 'Fissures in the Federal Structure? The Case of the University of Wales', *Higher Education Quarterly*, vol. 48, pp. 256-276.

Reader, W.R. and Hammond, N.V. (1994), 'A comparison of structured and unstructured knowledge mapping tools in psychology teaching', in N. Hammond and A. Trapp (eds.), *Proceedings of CiP 94*, University of York.

Richardson, J.T.E. (1995), 'Mature students in Higher Education II: An Investigation of Approaches to Studying Academic Performance', *Studies in Higher Education*, vol. 30, pp. 5-17.

Rogoff, A. (1999), *The Inside Careers Guide to Logistics Management 2000*, Inside Careers, Cambridge Market Intelligence Ltd., London, UK.

Rudzki, R.E.J. (1995), 'The Application of a Strategic Management Model to the Internationalisation of Higher Education Institutions', *Higher Education*, 29, pp. 421-441.

Smart, R. and Peterson, C. (1997), 'Super's career stages and the decision to change careers', *Journal of Vocational Behaviour*, vol. 51, pp. 358-374.

Spicer, D.P. (1998), 'Linking mental models and cognitive maps as an aid to organisational learning', *Career Development International*, vol. 3, issue 3, pp. 125-132.

Stevens, P. (1996), 'What works and what does not in career development programmes', *Career Development International*, vol. 1, issue 1, pp. 11-18.

St. John, E.P. and Andrieu S.C. (1995), 'The influence of price subsidies on within-year persistence by graduate students', *Higher Education*, vol. 29, pp. 143-168.

Stumpf, S.A. (1998), 'Corporate universities of the future', *Career Development International*, vol. 3, issue 5, pp. 206-211.

Super, D.E. (1990), 'A life-span, life-space approach to career development', in D. Brown and L. Brooks (eds.), *Career Choice and Development* (2nd edition), Jossey-Bass, San Francisco.

Teichler, U. and Carlson, J. (1990), 'Impacts of Study Abroad Programmes on Students and Graduates', *Higher Education Policy Series*, no. 11, vol. 2, Kingsley, London, UK.

Teichler, U. and Maiworm, F. (1994), 'Transition to work: the experiences of former ERASMUS students', *Higher Education Policy Series 28*, ERASMUS Monograph No. 18.

Thomas, B.J. (1995), 'Manpower and employment issues', *Maritime Policy and Management*, vol. 22, no. 3, pp. 239-253.

Tight, M. (1992), 'Part time Postgraduate Study in the Social Sciences: Students' Costs and Sources of Finance', *Studies in Higher Education*, 17, pp. 317-335.

Transport Planning Society (1997), *Abbreviated Statutes*, Transport Planning Society, London.

Tsakos, P.N. (1996), 'Shipping beyond 2000 - The views of an independent shipowner', the Reginald Grout Memorial Lecture, 1 May 1996, *Proceedings of the Chartered Institute of Transport*, vol. 5, no. 3, pp. 3-14, esp. p. 7.

Wang, S. (1996), 'A dynamic perspective of differences between cognitive maps', *Journal of the Operational Research Society*, vol. 47, pp. 538-549.

Webb, A. (1996), 'The expatriate experience: implications for career success', *Career Development International*, vol. 1, issue 5, pp. 38-44.

Wilcoxon, F. (1945), 'Individual comparisons by ranking methods', *Biometrics*, vol. 1, pp. 80-83.

Williams, G.L. (1992), *Changing Patterns of Finance in Higher Education*, Open University, Buckingham, UK.

Wilson, B. (1990), *Systems: Concepts, Methodologies and Applications*, 2nd edition, Wiley.

Wolstenholme, E.F. (1993), 'A case study in community care using systems thinking', *Journal of the Operational Research Society*, vol. 44, pp. 925-934.

Woodhall, M. (1989), 'Marketing British Higher Education Overseas: The Response to the Introduction of Full-cost Fees', *Higher Education Quarterly*, vol. 43, pp. 142-159.

Wright, P. (1996), 'Mass Higher Education and the Search for Standards: Reflections on some Issues Emerging from the "Graduate Standards Programme"', *Higher Education Quarterly*, vol. 50, pp. 71-85.

APPENDICES

Appendix 1.1 A Summary of a Transcript of Discussions of the Decision to Undertake Postgraduate Study in Shipping and Logistics at Plymouth

The discussion was held on 25 November 1996, with six members of the Diploma in Professional Studies in International Shipping and Logistics Management group, and lasted for 75 minutes.

All participants agreed to discussions being taped and anonymous recording of quotations / summaries in future publication. The summary of the transcript was agreed with all the participants soon after the interview was recorded. Their names have been recorded as Mr A, Mr B... Mr F. Their nationalities have been restricted to a continental region, and where possible the discussion is presented in the first person.

Mr A I am of North European origin, and have been at sea for two years, involved in shipping operations, and have spent four years at Nautical College. I had a long term aim of working ashore in brokering or chartering, rather than staying at sea for the whole of my career. I funded my studies three ways including a grant from the government, a loan and private funding for my tuition fees. [*Mr C* then commented that he was funding his studies in the same way.] I had eventually reached the point where the lure of travel and the excitement of seagoing work were no longer important to me and I had regarded the Masters course as the best option available to me at the time of applying, as a means towards obtaining further academic qualifications.

Mr B I am of South Asian origin and have spent seven years at sea involved in shipping operations. It was essential for me to gain postgraduate qualifications in order to obtain shore based work. [He had begun the initial discussion.] The University at Plymouth was the only institution offering me a course that had given me the opportunity to gain this qualification.

Mr C I am of North European origin, and have spent four years at sea involved in shipping operations. Eventually the pay cheque [Mr F 'the money'] becomes insufficient to compensate for the boredom of shipboard life. I am not yet too bored at this point in time, but probably will be considering my life time employment very soon. The ship's master has become a servant, merely responding to faxes and telexes, involved for example, in juggling with position telegrams in order to carry out crew changes and bunkering in port, with none of the power the job had once had years ago. I had gone to sea by accident. Initially I had planned to become a telecommunications engineer, but had accepted an offer from the Maritime College and Academy. (Why should I not do so?). I had always been fascinated, by ships and maritime operations. Now, I had reached the stage in my career where I felt that I needed to study to gain a broader understanding of shipbroking and chartering functions.

Mr D I am of East Asian origin, and have worked in logistics terminal operations. Whilst I am studying, I receive funding from my employers, the national government, who are also paying me a salary while I am attending the course. I am living with my family in Plymouth, and we all waited until after arriving here before finding our accommodation. My employers are paying for my living expenses and my tuition and other fees.

Mr E I am of Middle Eastern origin, and I am paying my own university fees, which at the full overseas rate, are more than double those of European Union students. I have had to save up for several years whilst working at sea. I had also hoped to go ashore and get married. I have worked in the military navy for two and a half years before then moving into merchant navy operations. The Centre for Maritime Studies at my previous university in the United Kingdom had guided me to contact the University of Plymouth, as I had applied first of all to that course, but it was an open learning programme. I had telephoned Plymouth, and was sent a course brochure. During this stage of the process, I was in London. Later I had an interview with the British Council (which had proved to be an important element in my decision) when I arrived in my home port.

Mr F I am of South European origin. I am very concerned about the differences between two different lives that I had had to lead, including the shore based life with my family and the mundane but stressful life aboard ship, away from my family. I had hoped to be able to remove the distinction between these two different lives. Altogether I had had 15 years of seagoing experience including four years during which I had undergone naval officer training. I had then felt that it was time for me to get involved in shore based work after the course, which should allow me to broaden my knowledge and understanding of shore based work.

[From this point onwards the discussion is paraphrased.]

Why Shipping?

Mr A: international business and shipping were always interesting.
Mr B: the excitement of travel had now worn off.
Mr C: initially the excitement of the job had been attractive.
Mr D: was interested in logistics.
Mr E: was 'always interested' in shipping, but he now wished to get married and to go ashore. He had studied in Naval Academy for four years, before joining the military navy. After two and a half years, he had resigned, and there was nothing left to do but to go to sea in the merchant navy. However, he liked his job.
Mr F: it was in his blood, with his relatives also at sea and having some influence on him. However, he now wanted to get ashore to spend time with his family.

Other Courses

Another British University had appeared to be attractive to *Mr A* and to *Mr E*, but the attraction of the 'postgraduate' status offered by Plymouth, rather than simply 'graduate' status at the other university, had been important to them. Also Plymouth was seen as being a 'harder teaching' course [*Mr D*]. A maritime university in another part of Europe was felt to be more expensive than the University of Plymouth.

Why Plymouth?

The opportunity to achieve a Master of Science degree (which Plymouth offered) was a very important issue to all concerned. *Mr C* pointed out that other shorter courses could be studied in his own country, which would eventually cover much of the material studied in the course at Plymouth, but without either giving a formal qualification or being available in a 'single broad course'.

Two North Europeans, had both been influenced by a mutual friend: *Mr A* had known him for 10 years; *Mr C* had met him in the navy eight years ago, but then had visited him and talked to one of the lecturers at Plymouth last year.
Mr B had been influenced by a Plymouth graduate.
Mr D: A lecturer, who had recently been on an exchange visiting professorship in Plymouth, and had been his lecturer at Maritime College in

his home country, had advised him that Plymouth was a 'beautiful location' and a good course.

Mr E: A leaflet and a discussion with the Maritime Studies department of his former university in the United Kingdom, and a friend who had studied in at another UK university last year, influenced him. The British Council in his home country had been important to him. The staff at Plymouth who were teaching him were felt to be very well qualified. Plymouth offers the opportunity for students seeking a successful career in the industry.

Mr F: had had friends on the course, who recommended it to him.

Barriers to Studying at Plymouth

Money

Mr B: Could have studied in the USA: two years in Canada plus one year in the United States would cost him the same as at Plymouth for a Masters degree, but would have taken a year longer, and time was very important to him.

Mr B pointed out that money was a real barrier [his words] with lots of his friends potentially interested in Plymouth, but unable to afford the fees which were charged.

Mr E: Many students from his home country would have studied in the UK if the fee were more attractive, similar to levels charged to students resident in the European Union.

Gaining a Place / 'Being Accepted'

Mr F was very anxious whilst waiting for the results of his application and was 'over the moon' when he eventually received his letter of acceptance.

Mr A and *Mr B* had also been anxious to receive their letter of acceptance.

Accommodation

Mr C and *Mr A* [as also had *Mr B*] had both been warned to avoid accommodation in the university halls of residence, before making their applications, because of the high cost and limited space of such accommodation. They had arrived in England to begin the course before finding and sharing a flat to live in. They were surprised to find living costs similar to those in their home country, having expected costs to be lower.

Mr E was living in the nearby university halls of residence, having taken up an offer of a place at Plymouth, whilst on shore leave in the summer holidays (July). However he had wished he had known about the accommodation arrangements earlier, as he might have decided to make alternative arrangements.

Mr D and *Mr F* had both found their own accommodation after arriving in Plymouth.

The Downside of the Plymouth Course?

Teaching Styles

Mr B and Norwegians expected a few examinations: coursework assignments took up a large part of their time possibly not allowing enough time for studying other issues.

Course Content

All present had recorded practical experience, as had the whole class of which they were a part, but they were very concerned about the basic nature of much of the content of the first semester. In particular, this related to the joint teaching with the French that included some younger and less experienced students. Some of them whom knew almost nothing practical [that the Keel is wet (*Mr A*), the bow is at the front, etc.]. *Mr A* found it insulting to be told that 'a container ship carries containers' when he had a nautical background. Practical experience should be recognised as having greater weight in the entry requirements, and he felt that he should get direct entrance to the Masters programme, and he did not understand why he had not been accepted directly on to that programme.

Mr B and the North Europeans expected more lectures on shipbroking, chartering and other 'practical' aspects of the course.

Other Comments

Other comments made related to high course fees that were an issue that might affect the attractiveness of Plymouth and practical seagoing experience was considered to be important to those concerned, and this should be more fully recognised in the admissions policies.

Mr E hadn't been sufficiently qualified to study for the Masters degree although he had studied in Naval Academy (after high school) for four years and had had seagoing experience for five years. He would have studied for the Masters degree in his own country, or at another UK university. Was this situation peculiar to University of Plymouth?

Appendix 1.2
A Transcript of a Focus Group on the Decision to Undertake Masters Level Study in Shipping and Logistics at Plymouth

The discussion was held on 3 December 1996, with six members of the Masters courses in International Shipping and International Logistics groups, and lasted for 80 minutes. Consent and abbreviations are as in Appendix 1. However, this is a verbatim transcription.

Ms A I am a merchant navy deck officer, and I have been teaching at a Naval Academy in my country in Western Europe for four years and I have now chosen to study for my Masters degree in International Logistics.

Mr B I am from Southern Europe and I am studying for my Masters degree in International Shipping.

Mr C I am from South Asia, and I work for a shipping corporation as a merchant sailor in my home country, and I am studying International Shipping.

Ms D I am from Central Europe and I have been taking a law degree. I am studying International Shipping.

Mr E I am from Southern Europe and I started my career as a seafarer. I have done a degree in ship competency in my home country, and I have two and a half years of sea going experience. I studied for a Diploma at Plymouth three years ago, and I have worked in a shipbroking company in my home country for the last three years and now I am back in Plymouth for my Masters degree in International Shipping.

Mr F I am from the Middle East, and following a Bachelors degree in Maritime Business, I worked in a contract company registered department for two and a half years. I am now in Plymouth for my Masters degree in International Shipping.

What was your Main Reason for Coming to Plymouth?

Mr E I came here because I studied for my Diploma here, but I first came because my institution knew that the universities that do shipping are limited: Plymouth and one other in the UK, and maybe one in Sweden and so on. So when you apply, you takc leaflets and so on, you count your money, and you say, I can go there. Actually because the information I got on paper was not so good, I didn't come here to live in Plymouth. I came here finally, because of the people.

Was this at Bachelors or Diploma level to start with?
Mr E I said it was the Diploma. I did it in 1992, before I went back to work again.

And what sort of work were you doing?
Mr E I was in shipbroking, a big company that was independent, and had both chartering and Sale and Purchase departments, but I was working in chartering.

So why did you chose Plymouth at this stage rather than this year or next year?
Mr E For me personally it had to do with my age. I am 28 so if I want to do it I have to do it now. I feel now that I still have the ability to concentrate and work at an academic level, and also because I want to enact now my future plans, and also financially, now that I have the money, to spend in doing my studies here. So, mainly these three reasons.

So what about your future plans ? What sort of work would you like to do?
Mr E I was thinking of working with shipping companies and also with business, shipping business, but now I find more fascinating an academic career in my home country or wherever.

You want to teach then?
Mr E Yes.

What sort of level do want to teach at?
Mr E Degree level.

In shipping?
Mr E Yes, or Nautical Studies, if I could do something about the outside world.

So how does the Plymouth course help you in that aim ?

Mr E The first thing is that you get the licence - lets say the ticket - the paper - which is the most important - [laughter] - when you go away from here you say I have a Masters, an MSc.

Is that true for everybody?

All Yes.

Mr E Nobody is going to check me for my language.

Mr B Precisely.

[laughter]

Ms A It is valid right across the world... you don't have to show your knowledge.

When you come to Britain and say that you studied in a British Institution in London or wherever, the fact is that when you show them an academic qualification from Britain, and that they accepted you to come to this University for a Masters, then, people see you differently. Somehow it is a point of reference that you have all over the world... and this is the all important thing from the course. When you decide to do a Masters it is essential to get it right because you will only do it once, you are not going to do it again. So you have to think very carefully whether it is worth studying in your own country which takes absolutely years or whether you should earn the money, come abroad and do it in the British Isles. Most people choose Plymouth because it is the only university in the world that I know that teaches International Logistics, and then you have to choose a university that has a course, and a tradition in that course, so that you can grasp the best knowledge there. I am not just here for doing it, I am here also for teaching. I have been teaching A levels, so that with a Masters degree I can teach in universities if I want to, because they are starting to develop these courses in my country. If I have a Masters I can find employment and they are starting courses in my home city in the next few years. In my home country, there is one person I know who has opened a logistics company, and I know him and he knows that I am here. And there are, besides, other people working in shipping who have all their life since their Masters so they know much about it. And there is a colleague of mine that went to a British university to teach logistics. In my country, he and I are the only people who have a degree in logistics. It gives me a wide scope for opportunity and the fact that I have studied logistics and the ICS course...

I've started, in fact I've taken the shipping option, but I don't know how I am going to cope with their courses, because they are very strict, but anyway this is my choice, because I can go back home and go to London with my qualification. I can only do it once, I cannot do it twice, and I think that logistics is what I want.

Did your government help you come to Plymouth then?

Ms A I am paying for it, I had to save the money and I didn't expect to come to Plymouth because I started my life in shipbroking and needed to do some course in shipping but I didn't want to leave home. I wanted to study at home, but then one of my lecturers told me that I had to come to Plymouth and that I should go for logistics in the present business environment, and I think that it was a good option that I took.

This discussion - was it a long time ago?

Ms A No, it was my lecturer who told me that I should go to Plymouth. He said don't go anywhere else, because he knew of the tradition and he knew what was going on here. Then suddenly I decided myself - it was August 1995 - I am going to Plymouth, because the other UK course is relatively new, you want to go to maritime school, and he started saying to me you must go to Plymouth. Do this, and you are young, and perhaps when you are older you can not do it, and would I be happy if I stayed at home and I did not do it? So I said, I am happy and I am going for it, and he started giving me the address. After the course I can still go into academia. I can try for management, because I will learn about decision making. If you have a Masters you can go for an academic university and you can cope with it well because you have some advanced courses. I think its good...

What about other people? What sort of ... is any of this familiar?

Mr B My brothers were in other universities in Britain, doing postgraduate courses, and I heard from them some concerns concerning the reliability of the British academic system, which gave proof of study without substitute. After that I did the Diploma at Plymouth, and as I was not satisfied, I came for the Masters. I chose the university very carefully from all the universities in England only. The United States is too far away, and other maritime countries are at that level, and my home country is not as good even though shipping is one-third of the income of there ...that's why I selected Plymouth university.

Why not another UK university?

Mr B Because it is not just shipping. It has some modules, but it is not pure shipping. It has more transport, but Plymouth is pure shipping and that is what I want. I intend to be specialised only.

Were there any family or friends or anybody who influenced your decision?

Mr B My father influenced my decision as my brother's decision and he tried to do this. At one point where I was very inspired and I decided to do maritime studies, shipping, the business of shipping. I chose that.

Is he paying or are you paying?

Mr B He's paying. This is usual in my country, your parents helping you, until you are finished with the army, your course, and you are established.

Have you been through the army yet?

Mr B I went through the army and after that I was free to choose with clear mind what I wanted to do.

So what are you going to do after the course at Plymouth? Did you have a job in mind?

Mr B What I will do is work for my father and my brothers in the family business concerning chartering or owning of ships ...I think the standards in Plymouth have a good value of time.

Time well spent?

Mr B Yes exactly.

What about you?

Mr C I studied a Diploma at a North European shipping academy and I got a distinction and I got the highest mark from this course and they recommended me for the course in Plymouth and they have written for course details for me. Actually we have a limited resources in our country at the academy of shipping and I think this is a good time for me, and I have to work another six years for my company who have given me finance for coming here.

So why Plymouth?

Mr C We have rather neglected Plymouth. In Britain, every university but London, Oxford, they don't care about Plymouth over there, as long as it is an English school.

Are there other people from your company who have been down this route?

Every year also there are people from the academy who come to shipping courses in Britain.

So when was that then... last year or some time ago?

Mr C No, last year two boys came. They got a distinction from their Masters and there were two the year before.

So Plymouth is good enough for the distinction candidate?
[laughter]

How about you?

Ms D Two questions. I had time to write, why go to this country and then why go to Plymouth? At the beginning I thought of looking at my languages, so I wanted to go abroad, so I looked round Spain, the UK and the US. Then I looked at the institutions and then I came up with Plymouth, and two other UK universities, and then as I wanted to go into shipping I thought Plymouth is the place to go because they concentrate on shipping and they've a very good reputation. It took me some time to find out about it because I'm from the inland southern part of Central Europe, and people in my home always look at their own qualifications. I know some people who've been to another British university, who know the university quite well, and they told me about the general way they have been treated and it did not really encourage me to go there.

Was there anything specific about that... the teaching styles or anything?

Ms D Well... equipment, all these computers and then the classes had vast numbers. My sister, she's been up at the other university and she told me of huge numbers of people in her class, so I wanted fewer people in the class, good qualifications. At that moment I felt that everybody was really encouraging me to go into the area that I wanted to go into and I felt that if this is what I am planning to go into, then you need this and this and this. Another thing is money. You don't want to do it when you are 62.

[laughter]

So why this year and not next year? How about you?

Mr F I searched all Europe's universities about shipping, I had applied too to the other UK university.

Why do you prefer England? There is America...

Mr F For a specialist, shipping subject, Britain is best. I applied to...

Sorry... How do you mean Britain is best?

Mr F I searched American universities, a north European university; a lot of universities do some shipping subjects, but not exclusively. Only British universities are best in this subject.

I applied to two universities, at Plymouth and elsewhere in the UK. I came here at the beginning of July, and I went to the other university, and I talked to last year's students, from Southern Europe and the Middle East.

You were talking to the students at the other university?

Mr F They weren't happy with their courses. Crowded lectures, many people... and the lectures didn't interest them. There were many problems. I came here and I talked to the students, and the lecturers made them welcome.

Okay , so it was talking to people that was important in your decision?

Mr F Yes. It's the best way, because in one subject or the other subject, the other way ...they send a catalogue, here's this subject and here's that subject. We are best in this subject ...they say here are our educational and other facilities, ...but I think its the best way.

So you think talking to last year's students is important? Was that the same for everybody? Did everybody talk to former students?

Ms D I would have liked to but I did not have the chance, because in my country nobody comes to the UK, but personally I think this is the best way to find out whether a course is worthwhile, from people who have attended it.

What about the importance of friends and people in Plymouth? Who actually knew somebody in Plymouth?

Ms A My lecturer back home was here and there was another boy from my country here. I will say...was here... he was taking his Masters in Oceanography. So when I came I phoned him and somehow he helped me and helped me to get around Plymouth, for the first days, and he was also finishing his dissertation, because he had to hand it in on the 30th of September and ...it was a good help.

Was the fact that somebody was actually here important in your decision? Did that affect your decision, or did that just make it easier once you had decided, or was that...

Ms A Err... no. The fact that my lecturer was here, or other people. The fact that he told me to come here, although he told me... It was my decision and I had to take it... As we were saying, you do it at one time in your life.

Mr E Yes that's right...

Ms A and err... as he was saying he was 28 and that's the right time... you do it at one time. I am older than him, so it is at this point that I have to do it... It is a matter of... it is a matter of at this moment in your life you are going to do it... There may come up opportunities and you have to have the right qualifications ...and I think that in the area of ports, things are going to change in the next two or three years. Then if I have the qualifications that nobody has, then because nobody has... One of my teachers, who was the leader of my course last year, was telling me (I am doing the ICS)... He has told me that when I come back from Plymouth I will be the most qualified person in the whole of my country that has these qualifications in shipping... [laughter] It's an opportunity that I have got to grasp...

What about the rest of you in relation to the people who you actually knew in Plymouth? Was it important to you in your decision to...

Mr C One person in my company, back in 1992 had been in Plymouth, and he knew all the places, and he was the one who told me about Plymouth and the city, and the nice life here, and he err...

So he was telling you about life in Plymouth?...

Yes...

Was that before you actually decided to come to Plymouth or afterwards? Was it important to you in the decision to apply to come here, or was it just useful information to you, after you had already decided to apply?

Mr C Yes I decided to apply.

Was it ...in terms of what he told you about Plymouth... was it that which made you want to apply?

Mr C No... because in our North European academy, he was the one who selected me to apply.

So he in fact almost did the applying for you? ...

Mr C Yes.

Ms D Yes I think its nice if you know somebody in the town, but if you consider something like education, then you really go for the course which you think is most interesting, and you know everything works out fine when you arrive. You know you're in... Plymouth, it's a city, so, well, I was never concerned about going to a place where I didn't know anybody. I mean, I've been abroad in many areas of the world...

So it was the course then, and not the city? Is that true for everybody?

Mr B When I came here, I came here without asking anybody but when I came here I saw a lot of academic disciplines that I knew from my home country...

When you came here, was that when you came here for an interview or when ...

Mr B No it was when I came here in October. I was determined to come... all my sisters had finished. I was finishing...

So this was before the course actually started that you came to Plymouth then?

Mr B I wanted to come here, and I was determined. I made my application from home and I came...

So that's two of you who knew nobody? Is that right? Is there anybody else who knew nobody at all?

Mr E When I was doing my degree at the Maritime Academy back home, I finished in 1990, I knew from my years at sea, that it was part of my plan to gain further qualifications. So I knew this. Go to Britain. It was a must, not a random decision.

Why Britain and why not America?

Mr E When you have been in your maritime academy for three years, and you hear all your lecturers saying, if you want to go further, you must go there, that's it I mean. When you are 19, and your lecturer says you have to go there to gain the qualification, it goes in your mind, and you have to go. There is no other way.

Ms A It is a tradition...

Mr F Yes that's right, it's a tradition. You speak to people and...

Ms A My work mate... he didn't tell me to go to a North European university... he told me to go to Plymouth.

Mr E It has a worldwide reputation.

Ms A People say... don't go to the other UK university ... I didn't know if it had a shipping course. People just told me, you have to go to Plymouth.

So even if you didn't know the people, its the personal reputation that matters?

All Yes [times 5].

Ms D Reputation is what everybody has, its err... the best place in the world. If you have to go to a Masters, then you have to go to Plymouth. You must go there, and I think its the best in Europe, its the best in the rest of the world. It is the right place you have to go... you have to go there... you must go there...

Mr E So I have been deciding I will come to Britain... so I start searching and so on... but in 1991 I met the partner I am with now, but she's English. We met in Southern Europe five years ago, and she's living in Looe, her father's house... its close to Plymouth. I could have gone to the other UK university, the distance is not that significant, but who knows, it was the reason for first coming to Plymouth, and of course now, for living in Plymouth. I am of course here now for the second time... it is not exactly information from inside the system, but it influenced me, to come here to Plymouth.

Ms D Could I just say something about the country issue here, because before you raised the issue of why not go to the United States, I think that particularly in shipping it is quite obvious to go the UK; it's a traditional thing. ...A country of seafarers, and when I then later on told people, yeah I was doing shipping in the UK, everybody said, shipping in the UK, yeah, sure... country of seafarers, so I think there is something traditional there.

Mr B The academic system in Britain is a very good industry and in relation with the conservative natural Briton, in our eyes, in my friends' eyes at least, makes the British academic system, very strong, at the top. We don't go to Southern Europe, for example, to study shipping, or anywhere else... it's not only shipping. The British system combines all the elements. You can do everything, if you want to do engineering, from law to engineering.

Ms A And one more thing... I am very sorry for interrupting... I am a student member of the ICS... and I did some reading from Maritime Policy and Management. Yesterday I was reading through one of the magazines and I came across one of the articles by the people that are by my teachers, Dr X and Professor Y. You come across all these things that go into these articles in these magazines, that are the best ones in the world. You are glad that you are being taught by them, because they are recognisable all over the world, and if you want to take a Masters then you must do it with the right people at the right place. It is Plymouth.

Mr C Yes.

What about accommodation? Did that affect your decision about coming to Plymouth, or applying for Plymouth?

Ms D I wanted to go there and then things will work out fine.

Was it an issue for anybody?

No.

OK, so what about fees and things like that?

[general laughter...]

Ms D Being a member of the European Union, its not too bad.

Mr F I'm not. I'm self-financing. I paid the overseas fee rate.

How much?

Mr F £X.

That's a lot of money isn't it?

Mr F It certainly is.

Who's paying for that then?

Mr F I and my parents.

Right, so who made the decision about Plymouth? You or... ?
Mr F Yes.

So it was you?
Mr F Yes.

So what did your parents say about coming to Plymouth? Did they influence things...?
Mr F No. No, they just pay...
[general laughter]
Mr C I've been working and could come here definitely myself ...but they just decide to give me the money... so it helped... it has happened...

So what did you do? I mean were the fees an issue at Plymouth? If you were paying all this money...
Mr F No, but you are more expensive than the other university; you are a few hundred pounds more here.

Was money important to you?
Mr F No... well, yes. Money is always important, yes.
Ms D Don't say, because they will put the fees up.
Mr F I am reversing what I was saying yes?

To an extent the price did influence your decision?
Mr F ...I had just two choices... the other UK university and Plymouth... I preferred Plymouth... money came second...

Alright, but if it had been a lot more, the other UK university might have been more attractive?
I would have gone to Cardiff had it been more than it was...

OK what about anybody else? Are fees in the EU an issue?
Ms D I came from a country where education is totally free and I find the situation where I had to pay fees which were appallingly high... so I thought well, ...but that's the way it is...
Ms A In my country, being in the EU, the fees are very cheap in relation to the fees I would pay in my home country. It takes three years almost to take a Masters there, and there you pay more. They opened the postgraduate course, this year, and they are paying more money than I am paying here. Anyway, I am a member of the European Community and I have that advantage and that must be taken into account, because if not, then I will pay much more and I will have to be very careful.

Ms D I think there is something else. I think had I not come from the EU I would not have chosen Plymouth. The overseas fee rate is an awful lot of money. I think for this money I would probably have gone to the United States.

But given that where you were...

Ms D This way its OK.

What about teaching methods? Was this an issue when you were applying? Did you know anything about what to expect at Plymouth? Was it important for anybody about teaching methods?

Mr B Since I was coming in England, I knew that they were very good, and if I wasn't satisfied with my work, I shouldn't think of it, because after all it was not the British method, but I should change my methods and conform to that.

[laughter]

Ms D Well, what I did, the British Council, they have a clever book, listing all the universities with the courses and they give numbers, average numbers of students in class, way of teaching, assessment and so on. I looked at this book first, and I thought, it looks good, and then I wrote a letter actually to the faculty, saying, give me average students numbers and so on. Because I think this to be very important, and I found it to be very good, and then I send it to them.

Was that before you applied or after you applied?

Ms D This was at the same time, because I did not have a lot of time. But I am sure that had I got some information that I considered not to be very good, then I would probably have not chosen Plymouth.

So that was important to you?

Ms D Yes, very.

What about other things... what have we not talked about in the decision to come to Plymouth or to study?

Mr E The future, I think.

OK, Would you like to tell me something about that?

Mr E Yes. If people are satisfied with what they are doing, then they don't go for further education. If someone has worries and wants to make a step forward, they start reassessing, then they start looking around. The academic system in Britain, is very, very good, it is closer related to industry than my home universities. It is not so art, so academic education, the people who come out are a bit more relevant. There is still a great gap, but it is still more relevant, closer to the deep waters.

Ms A Teaching technique can be very tough, in universities in my country. Problems with mathematics and physics and chemistry that students have got, depending on the courses they take. They have three years of mathematics in a five-year course, even at Masters level they teach you Mathematics, but one of the things about your courses is that the Mathematics at your course is good. Not only at Masters level, but also at undergraduate level, because I have a student of mine at another British University taking naval architecture, and small boats. He told me that the mathematics he had taken at our school was enough for the three years without worrying about it. Which is good because the difficult things you turn into very simple and practical things. If you want to learn the background of it, then you can do so because you've got these things available. But, you don't have the need to know the formula from the beginning to the end which is important because you just have to know the end, though you've got to know the conclusions, then you can go and investigate. When you know the theoretical part of it and you know the practical part of it, that is a very good thing because I've had statistics last year.

And Plymouth makes you use both of those?

Ms A I had statistics last year and I had this mathematical teacher, and she just read to the class with all the formulas on the board, and then she just gave us this exercise and she didn't tell us what she was saying, which is to take conclusions and analyse statistics. Then when I came here, I came to start to understand and learn to take conclusions from the values that I get from statistics. This is very important because as soon as you get practical, this is very important.

Is there anything else which we haven't talked about?

Mr F ...concerns about chartering and brokering. Not relevant to decisions...

Mr B Structure of the course is not really relevant to the decision to study at Plymouth... but what counts more is the Plymouth history of that particular university, influenced by others and friends and...

Its opinions really?

Mr C Yes - it depends on where you want to go. We want the best. Now that we are paying, we want the best...

Ms D You pay only once for a Masters course, so if you are paying, you want to pay for the best. I never heard of thinking of the United States for shipping. You know the US for shipping doesn't sound good. One doesn't link shipping with the United States.

Mr E We are good Europeans, so we think differently.

Mr C Exactly. But I go to England for shipping. I wouldn't go to other countries in Northern or Western Europe, because they don't have like 500 years of tradition in shipping.

Mr B The Southern European community in shipping gives so much support for England... why?... because it was based in England for so long.

Mr C The ICS course I am doing gives me a strong base within the industry. Their courses and exams are very tough. Their courses are recognised internationally for shipbroking. It is not an academic course but a professional qualification that is recognised and gives you the recognition that you need.

[The discussion ended, but was followed by a request to restart.]

Ms D Something that made me choose the UK, rather than going to other European countries, is that over here there is something like tutorship. I found that I arrived here two weeks late, and I came in and the course leader gave me help telling me we have been doing this and this, and he very practically helped me to find accommodation. At home this would have been unthinkable. There they just want your money and say 'See you on Monday'. Here I find everybody helpful. The idea of a personal tutor is a really good one.

Mr C I would agree with that. I made my application for studying here and I didn't receive anything at all for over a month. I was really going crazy, and I had to contact somebody. I had to contact my employer, to get a letter to allow me to go to England and I wasn't hearing from you, and I had to get that licence... but everything was OK. It was my personal fear... you know... because the things are so well organised that you needn't think about it, because all these people, all this organisation is perfect. I phoned the course leader and he said don't worry, its OK, you are just being processed with all the other students. I had my acceptance at the end of January. From January to July I was really worrying you know, but I phoned here but everything was OK and I had no problem.

Ms A Everything must be so organised because I could not think of applying for a course in my own country - it's impossible. I mean this institution has a really high reputation and I am surprised that lecturers here don't have a lot of student helpers doing their photocopying and so on. I am really surprised because I am used to working for my lecturers and professors and here I was surprised that academic staff have no assistants, no staff, and doors open. Lecturers have time to open the door and say yes, come in, to help you create some new ideas and work on projects...

That's good ... can I turn this machine off?

Appendix 2.1 Questionnaire for Plymouth Postgraduates

The Decision to Study Shipping or Logistics at Plymouth, at Postgraduate or Post-experience Level

John Dinwoodie, Room P10

Background

This work follows earlier studies of the undergraduate decision to study transport, and interests in careers issues in shipping and logistics. It supports teaching on the Business Systems module in Stage 2 of the Masters course.

Your name *Date*

All details are confidential, but please show your name so that I can discuss / follow-up replies.

Please answer ALL the questions below, but you are, of course, free to withdraw from this exercise at any point, if you so wish. Please write any other comments / reactions at any point in the questionnaire. Please state any objections that you would have about me using your answers anonymously in an appendix to a thesis, or other published material.

Comment:

Please describe your current course of study.

Ring one of each which applies:

Level: MSc / Diploma *Period / Stage* 1 2 3
Subject: Shipping / Logistics / Both

Please think of yourself in relation to postgraduate or post-experience or Masters level study in shipping or logistics at Plymouth.

What factors were important to you in this context?

Preliminary discussions identified the factors and items shown below.

If a particular factor was not an issue to you in this context, please tick the box provided and move on to the next factor.

If the factor was an issue, tick one box against each item rate its importance in your decision as critical / important / relevant / irrelevant.

If other items were relevant please describe them in the 'other: please specify' section.

1. How did your family or friends influence your decision to study? How important were they?

Critical Important Relevant Irrelevant	*Cr*	*Im*	*Re*	*Ir*
This factor was not an issue for me	☐	☐	☐	☐
My family circumstances are relevant	☐	☐	☐	☐
They introduced me into the industry	☐	☐	☐	☐
I talked to them	☐	☐	☐	☐
They influenced me because they are funding my studies	☐	☐	☐	☐
Other: please specify	☐	☐	☐	☐

2. What reasons made studying logistics / shipping at postgraduate or post-experience level attractive? How important was each?

Critical Important Relevant Irrelevant	*Cr*	*Im*	*Re*	*Ir*
This factor was not an issue for me	☐	☐	☐	☐
<u>I wanted</u>:				
to broaden my knowledge / learn new things	☐	☐	☐	☐
to enact my long term career plans	☐	☐	☐	☐
a change of career / to go ashore	☐	☐	☐	☐
to broaden my opportunities / be sure to find a job	☐	☐	☐	☐
to specialise in shipping as I want to work there	☐	☐	☐	☐
more study, following my work experience	☐	☐	☐	☐
Other: please specify	☐	☐	☐	☐

3. What reasons might have put you off studying logistics or shipping at postgraduate or post-experience level? How important was each?

Critical Important Relevant Irrelevant	*Cr*	*Im*	*Re*	*Ir*
This factor was not an issue for me	☐	☐	☐	☐
I did not yet have the money / was saving up, while at sea	☐	☐	☐	☐
I needed operational experience to improve my understanding	☐	☐	☐	☐
Had I experienced work or family business pressures	☐	☐	☐	☐
I had to be sure it was the right course as I could not repeat it	☐	☐	☐	☐
If I was offered a job giving good experience	☐	☐	☐	☐
Returning to study after ten years working would be hard	☐	☐	☐	☐
Other: please specify	☐	☐	☐	☐

4. How did you find out about courses in these areas? How important was each source?

Critical Important Relevant Irrelevant	*Cr*	*Im*	*Re*	*Ir*
This factor was not an issue for me	☐	☐	☐	☐
I talked to a work friend / people in industry	☐	☐	☐	☐
My previous lecturers told me about it	☐	☐	☐	☐
I contacted the British Council	☐	☐	☐	☐
I read magazines / books / brochures	☐	☐	☐	☐
By chance	☐	☐	☐	☐
I talked to others planning to study there	☐	☐	☐	☐
I talked to students / former students on the course	☐	☐	☐	☐
Other: please specify	☐	☐	☐	☐

in Shipping and Logistics

ogistics or shipping in the UK

	Cr	*Im*	*Re*	*Ir*
	☐	☐	☐	☐
resting	☐	☐	☐	☐
around the world	☐	☐	☐	☐
untry	☐	☐	☐	☐
hipping to gain	☐	☐	☐	☐
in less time than at	☐	☐	☐	☐
evant to industry	☐	☐	☐	☐
ipping tradition	☐	☐	☐	☐
	☐	☐	☐	☐

istics or shipping at Plymouth [or udents] attractive? How important

	Cr	*Im*	*Re*	*Ir*
	☐	☐	☐	☐
	☐	☐	☐	☐
hen looking for a job	☐	☐	☐	☐
	☐	☐	☐	☐
ics	☐	☐	☐	☐
	☐	☐	☐	☐
ry to this	☐	☐	☐	☐
c Stage 1	☐	☐	☐	☐
course	☐	☐	☐	☐
	☐	☐	☐	☐
city	☐	☐	☐	☐

For St Nazaire students only

<u>By choosing St Nazaire, its links with Plymouth gave me</u>:

a chance to study in Plymouth, and not just any UK city	☐	☐	☐	☐
a Plymouth degree, which is recognised and easier to find a job with	☐	☐	☐	☐
access to more study facilities	☐	☐	☐	☐
Other: please specify	☐	☐	☐	☐

7. What reasons might have put you off studying logistics or shipping at Plymouth? How important was each?

Critical Important Relevant Irrelevant	*Cr*	*Im*	*Re*	*Ir*
This factor was not an issue for me	☐	☐	☐	☐
<u>If I had found that Plymouth University</u>:				
gave me too little information	☐	☐	☐	☐
was costly in relation to other courses	☐	☐	☐	☐
rated below other universities on its course quality, library etc.	☐	☐	☐	☐
had not offered me a place at MSc level	☐	☐	☐	☐
qualifications were not considered unique by employers	☐	☐	☐	☐
considered my grades to be inadequate	☐	☐	☐	☐
was not the best school which I could afford	☐	☐	☐	☐
students had given me bad reports about it	☐	☐	☐	☐
Other: please specify	☐	☐	☐	☐

8. Think of courses that you could have studied elsewhere. Why did you not study elsewhere? How important was each reason?

Critical Important Relevant Irrelevant	*Cr*	*Im*	*Re*	*Ir*
This factor was not an issue for me	☐	☐	☐	☐
<u>Elsewhere</u>:				
it was a totally different subject	☐	☐	☐	☐
I would have had to repeat material I had already studied	☐	☐	☐	☐
the course emphasis was wrong for me	☐	☐	☐	☐
it was very expensive, and not necessarily better	☐	☐	☐	☐
the course took longer to complete	☐	☐	☐	☐
Other: please specify	☐	☐	☐	☐

in Shipping and Logistics

lowing barriers which might have ng at Plymouth?

	Cr	Im	Re	Ir
	☐	☐	☐	☐
	☐	☐	☐	☐
at I wanted	☐	☐	☐	☐
ered the total time	☐	☐	☐	☐
and encouraged me	☐	☐	☐	☐
	☐	☐	☐	☐
accommodation etc.	☐	☐	☐	☐
	☐	☐	☐	☐
eel welcome	☐	☐	☐	☐
d not merely to	☐	☐	☐	☐
ued in entry	☐	☐	☐	☐
its to companies	☐	☐	☐	☐
	☐	☐	☐	☐

n overlooked? How important was

t	Cr	Im	Re	Ir
	☐	☐	☐	☐

General

Are you male or female? (tick one) Male |_| Female |_|

What is your age? (tick one)
20 - 29 |_| 30 - 39 |_| 40 - 49 |_| 50 + |_|

What is your nationality?

Please describe the most relevant work experiences you have in relation to the course:

1. Job title Full / part time?

Name and city of organisation

Duration (years and months)

How has this influenced your views about postgraduate study in shipping and logistics?

2. Job title Full / part time?

Name and city of organisation

Duration (years and months)

How has this influenced your views about postgraduate study in shipping and logistics?

Have you any other comments?

Draw a 'mind map' linking each factor which was an issue by using arrows to show the direction of cause and effect.
Show how strong each link is by numbering them:
1 = weak 2 = moderate 3 = strong, with a positive sign for positive links, and a negative sign for negative links.

A diagram is included to assist you:

9. Barriers to study at Plymouth

8. Reasons for not studying elsewhere

7. Reasons for not studying L/S at Plymouth

6. Attractions of studying L/S at Plymouth

5. Attractions of studying L/S in the UK

4. Information about courses

3. Reasons for not undertaking pg. study in L/S

2. Attractions of pg. study L/S

1. Family or friends

Show how strong each link is numbering them:
1 = weak 2 = moderate 3 = strong.
+3 is a strong positive link (A affects B positively and strongly)
-2 is a moderate negative link (i.e. A affects B moderately and inversely).

Code	1	2	3	4	5	6	7	8	9
1. Family / friends									
2. Attractions of pg. study									
3. Not pg. study									
4. Course information									
5. Attractions of study in UK									
6. Attractions of study in Pl									
7. Why NOT study in Pl									
8. Why NOT study elsewhere									
9. Barriers to study at Pl									

Appendix 2.2 Questionnaire for Postgraduates Nationally in the UK

The Decision to Study Shipping or Logistics at Postgraduate or Post-experience Level

Background

We are investigating the decision to undertake postgraduate and post-experience courses in shipping and logistics at various universities.

Please answer ALL the questions below, but you are of course free to withdraw from this exercise at any point, if you so wish. Please add any other comments or reactions at any point in the questionnaire. Please make any comments that you would have about me using your answers anonymously in an appendix to a thesis, or other published material.

Comment:

Q0. Please describe your current course of study. (Please tick all relevant boxes.)
I am currently studying at the following university:

Cardiff	☐	Huddersfield	☐	Plymouth	☐
Central England	☐	City	☐	Salford	☐
Cranfield	☐	Guildhall	☐	Southampton	☐
East London	☐	Liverpool John Moores	☐	Other	☐

Level: Masters ☐ Diploma ☐ Other (please state)
Subject: Shipping ☐ Logistics ☐ Both ☐

Please think of yourself in relation to postgraduate or post-experience study in shipping or logistics at university.

What factors were important to you in this context?

Focus groups and pilot surveys identified the factors and items below.
Please tick the box provided to rate the importance of each factor as an issue to you in this context. If no item is relevant, move on to the next factor.
If a factor was an issue, tick one box against each item within it to rate its importance in your decision as *cr*itical, *im*portant, *re*levant or *ir*relevant.
If other items were relevant, describe them in the 'other: please specify' section.

1. How did your family or friends influence your decision to study? How important were they?

	Cr	*Im*	*Re*	*Ir*
This factor overall as an issue was:	☐	☐	☐	☐
My family circumstances are relevant	☐	☐	☐	☐
They introduced me into the industry	☐	☐	☐	☐
I talked to them	☐	☐	☐	☐
They influenced me because they are funding my studies	☐	☐	☐	☐
Other: please specify	☐	☐	☐	☐

2. What reasons made studying logistics or shipping at postgraduate or post-experience level attractive? How important was each?

	Cr	*Im*	*Re*	*Ir*
This factor overall as an issue was:	☐	☐	☐	☐
I wanted:				
to broaden my knowledge / learn new things	☐	☐	☐	☐
to enact my long term career plans	☐	☐	☐	☐
a change of career / to go ashore	☐	☐	☐	☐
to broaden my opportunities / be sure to find a job	☐	☐	☐	☐
to specialise in shipping as I want to work there	☐	☐	☐	☐
more study, following my work experience	☐	☐	☐	☐
Other: please specify	☐	☐	☐	☐

3. What reasons might have put you off studying logistics or shipping at postgraduate or post-experience level? How important was each?

	Cr	*Im*	*Re*	*Ir*
This factor overall as an issue was:	☐	☐	☐	☐
I did not yet have the money / was saving up, while working	☐	☐	☐	☐
I needed operational experience to improve my understanding	☐	☐	☐	☐
Had I experienced work or family business pressures	☐	☐	☐	☐
I had to be sure it was the right course: I could not repeat it	☐	☐	☐	☐
If I was offered a job giving good experience	☐	☐	☐	☐
Returning to study after ten years working would be hard	☐	☐	☐	☐
Other: please specify	☐	☐	☐	☐

4. How did you find out about courses in these areas? How important was each source?

	Cr	*Im*	*Re*	*Ir*
This factor overall as an issue was:	☐	☐	☐	☐
I talked to a work friend / people in industry	☐	☐	☐	☐
My previous lecturers told me about it	☐	☐	☐	☐
I contacted the British Council	☐	☐	☐	☐
I read magazines / books / brochures	☐	☐	☐	☐
By chance	☐	☐	☐	☐
I talked to others planning to study there	☐	☐	☐	☐
I talked to students / former students on the course	☐	☐	☐	☐
Other: please specify	☐	☐	☐	☐

5. What reasons made studying logistics or shipping in the UK attractive? How important was each?

	Cr	*Im*	*Re*	*Ir*
This factor overall as an issue was:	☐	☐	☐	☐
<u>I wanted</u>:				
to go abroad as it makes life more interesting	☐	☐	☐	☐
a British Diploma / MSc is recognised around the world	☐	☐	☐	☐
an MSc: we don't have them in my country	☐	☐	☐	☐
to practise English, the language of shipping, to gain more opportunities	☐	☐	☐	☐
a UK course which can be completed in less time than at home	☐	☐	☐	☐
an academic system which is more relevant to industry	☐	☐	☐	☐
to study in a land with 500 years of shipping tradition	☐	☐	☐	☐
Other: please specify	☐	☐	☐	☐

6. What reasons made studying logistics or shipping at your university attractive? How important was each?

	Cr	*Im*	*Re*	*Ir*
This factor overall as an issue was:	☐	☐	☐	☐
My university's reputation:				
was commended by my lecturers	☐	☐	☐	☐
and tradition is worldwide, important when looking for a job	☐	☐	☐	☐
The course:				
is the only one which offers an MSc	☐	☐	☐	☐
is the only one in international logistics	☐	☐	☐	☐
specialises in shipping	☐	☐	☐	☐
offers an MSc: the Diploma is an entry to this	☐	☐	☐	☐
could give me remission from the MSc Stage 1	☐	☐	☐	☐
In the city of my university:				
I have friends who have finished the course	☐	☐	☐	☐
I had already undertaken studies	☐	☐	☐	☐
I would live in a nice place / beautiful city	☐	☐	☐	☐
My university:				
has a good admissions administration	☐	☐	☐	☐
accepted me	☐	☐	☐	☐
did not have bad oral reports like other universities	☐	☐	☐	☐
lecturers have a good reputation	☐	☐	☐	☐
Other: please specify	☐	☐	☐	☐

7. What reasons might have put you off studying logistics or shipping at your university? How important was each?

	Cr	*Im*	*Re*	*Ir*
This factor overall as an issue was:	☐	☐	☐	☐
If I had found that my University:				
gave me too little information	☐	☐	☐	☐
was costly in relation to other courses	☐	☐	☐	☐
rated below other universities on its course quality, library etc.	☐	☐	☐	☐
had not offered me a place at MSc level	☐	☐	☐	☐
qualifications were not considered unique by employers	☐	☐	☐	☐
considered my grades to be inadequate	☐	☐	☐	☐
was not the best school which I could afford	☐	☐	☐	☐
students had given me bad reports about it	☐	☐	☐	☐
Other: please specify	☐	☐	☐	☐

8. Think of courses which you could have studied elsewhere. Why did you not study elsewhere? How important was each reason?

	Cr	*Im*	*Re*	*Ir*
This factor overall as an issue was:	☐	☐	☐	☐
Elsewhere:				
it was a totally different subject	☐	☐	☐	☐
I would have had to repeat material I had already studied	☐	☐	☐	☐
the course emphasis was wrong for me	☐	☐	☐	☐
it was very expensive, and not necessarily better	☐	☐	☐	☐
the course took longer to complete	☐	☐	☐	☐
Other: please specify	☐	☐	☐	☐

9. How important are each of the following barriers which might have put you off studying logistics or shipping at your university?

	Cr	*Im*	*Re*	*Ir*
This factor overall as an issue was:	☐	☐	☐	☐
9a. Money:				
This factor overall as an issue was:	☐	☐	☐	☐
I was willing to pay a bit more for what I wanted	☐	☐	☐	☐
The total cost was less, when I considered the total time involved	☐	☐	☐	☐
My company or parents partly funded and encouraged me	☐	☐	☐	☐
Other: please specify	☐	☐	☐	☐
9b. The quality of life in the city, my accommodation etc.	☐	☐	☐	☐
9c. Teaching methods on the course:				
This factor overall as an issue was:	☐	☐	☐	☐
My first contacts with staff made me feel welcome	☐	☐	☐	☐
I expected to understand and learn, and not merely to analyse statistics	☐	☐	☐	☐
My practical experience was undervalued in entry requirements	☐	☐	☐	☐
I expected a practical course, with visits to companies and professional links	☐	☐	☐	☐
Other: please specify	☐	☐	☐	☐

10. Are there any issues that have been overlooked? How important was each?

	Cr	*Im*	*Re*	*Ir*
Please specify	☐	☐	☐	☐
Please specify	☐	☐	☐	☐

General

Are you male or female? (tick one) Male ☐ Female ☐

What is your age? (tick one)
20 - 29 ☐ 30 - 39 ☐ 40 - 49 ☐ 50 + ☐

What is your nationality?

Please describe the most relevant work experiences you have in relation to the course:

1. Job title Full / part time?

Name and city of organisation

How has this influenced your views about postgraduate study in shipping and logistics?

2. Job title Full / part time?

Name and city of organisation

How has this influenced your views about postgraduate study in shipping and logistics?

Have you any other comments?

Thank you for your help. Please return the form now.
John Dinwoodie, IMS, University of Plymouth, Drake Circus, Plymouth, PL4 8AA

Appendix 2.3
National Questionnaire for Maritime Business Undergraduates

The Decision to Study Maritime Business at Undergraduate Level at British Universities

John Dinwoodie, IMS, University of Plymouth, Drake Circus, Plymouth Devon PL4 8AA

Background

We are investigating the decision to undertake undergraduate courses in maritime business and related subjects at various UK universities.

Please answer ALL the questions below, although you may withdraw from this exercise at any point, if you so wish. At a later date, it may be useful to quote from your answers anonymously, or publish overall totals. Please state any objections which you would have in this context.

Comment:

Please describe your current course of study.

Ring one of each which applies:

Level: BSc / HND / Other   *Year or stage:* 1 2 3
Course title:

Please think of yourself in relation to the decision to embark on undergraduate study in maritime business at a UK university. What factors were important to you in this context?

Preliminary surveys identified the factors and items shown below.

Tick one box against each item to rate it as *very*, *quite*, *ind*ifferent or *not* at all attractive or important etc.

1. At present, how attracted are you to each of the following careers in maritime business on graduation?

	Ve	*Qu*	*Ind*	*Not*
Deck officer	☐	☐	☐	☐
Marine Insurance	☐	☐	☐	☐
Ship broker	☐	☐	☐	☐
Ship manager	☐	☐	☐	☐
Port manager	☐	☐	☐	☐
Maritime lawyer	☐	☐	☐	☐
Transport manager	☐	☐	☐	☐
Freight forwarder	☐	☐	☐	☐
Importer / exporter	☐	☐	☐	☐
Marine leisure	☐	☐	☐	☐
Other: please specify	☐	☐	☐	☐

2. How would you find out about courses and careers in maritime business? How important is each source?

<u>I would</u>:	*Ve*	*Qu*	*Ind*	*Not*
talk to people in the industry	☐	☐	☐	☐
talk to my careers advisers	☐	☐	☐	☐
talk to my lecturers	☐	☐	☐	☐
read job advertisements	☐	☐	☐	☐
read magazines / books / brochures	☐	☐	☐	☐
write to companies direct	☐	☐	☐	☐
talk to friends / relatives	☐	☐	☐	☐
attend careers presentations	☐	☐	☐	☐
Other: please specify	☐	☐	☐	☐

3. What reasons made studying maritime business at university attractive? How important was each reason?

<u>I was attracted to study in this area because</u>:	*Ve*	*Qu*	*Ind*	*Not*
of my previous work experience	☐	☐	☐	☐
of my interest in the sea	☐	☐	☐	☐
of the course reputation	☐	☐	☐	☐
of advice from my previous school / course	☐	☐	☐	☐
I needed to improve my job prospects / begin a career	☐	☐	☐	☐
of parental pressure	☐	☐	☐	☐
my friends were doing the same	☐	☐	☐	☐
Other: please specify	☐	☐	☐	☐

4. I am currently studying at the following university. (Please tick one box only.)

Cardiff	☐	Southampton IHE	☐
Newcastle	☐	Humberside	☐
City	☐	Guildhall	☐
Liverpool John Moores	☐	Plymouth	☐

Other: please specify

5. What reasons attracted you to choose your present university as a centre at which to study maritime business? How important was each reason?

<u>My university</u>:	*Ve*	*Qu*	*Ind*	*Not*
is in a city with a good social life	☐	☐	☐	☐
is located near the sea	☐	☐	☐	☐
is located in a nice place	☐	☐	☐	☐
has a worldwide reputation	☐	☐	☐	☐
offers a course which is good / unique	☐	☐	☐	☐
is near home	☐	☐	☐	☐
seemed to offer good facilities	☐	☐	☐	☐
Other: please specify	☐	☐	☐	☐

6. What reasons might have put you off studying maritime business at your present university? How important was each factor?

Nothing could have put me off:	*Ve*	*Qu*	*Ind*	*Not*
If I had been offered a job before arriving at university	☐	☐	☐	☐
If I had felt it was too distant from home	☐	☐	☐	☐
If I had obtained poor grades, or felt unable to cope	☐	☐	☐	☐
If the course content or structure had been poor	☐	☐	☐	☐
If I had felt the local climate / weather was not good	☐	☐	☐	☐
If I had considered it to have poor resources	☐	☐	☐	☐
If I had not liked the atmosphere there	☐	☐	☐	☐
Other: please specify	☐	☐	☐	☐

7. Which one other university did you most carefully consider studying maritime business at? (Please tick one box only.)

Cardiff	☐	Southampton IHE	☐
Newcastle	☐	Humberside	☐
City	☐	Guildhall	☐
No other university	☐	Liverpool John Moores	☐
Plymouth	☐	Other: please specify	☐

8. Why did you consider studying maritime business other than at your present university? How important was each reason?

	Ve	*Qu*	*Ind*	*Not*
They offered me a place	☐	☐	☐	☐
I did not consider studying elsewhere				
They offered a better course	☐	☐	☐	☐
They offered a similar course	☐	☐	☐	☐
The other university was in a more exciting city	☐	☐	☐	☐
The other university was nearer home	☐	☐	☐	☐
Other: please specify	☐	☐	☐	☐

9. Are there any issues that have been overlooked? How important was each?

	Ve	*Qu*	*Ind*	*Not*
Other: please specify	☐	☐	☐	☐
Other: please specify	☐	☐	☐	☐

General

Are you male or female? (tick one) Male ☐ Female ☐

What is your age? (tick one)
20 - 29 ☐ 30 - 39 ☐ 40 - 49 ☐ 50 + ☐

What is your nationality?

Please tick one box to identify the single most relevant area of work experiences you have had in relation to the course.

Deck officer	☐
Ship manager	☐
Transport manager	☐
Marine leisure	☐
Marine insurance	☐
Port manager	☐
Freight forwarder	☐
Nothing relevant	☐
Ship broker	☐
Maritime lawyer	☐
Importer/exporter	☐
Other: please specify	☐

Was it full time or part time? (tick one box)
Full time ☐ Part time ☐ None ☐
How long were you employed? (tick one box)
None ☐ Under 1 year ☐ 1 - 3 years ☐ Over 3 years ☐
Have you any other comments?

Thank you for your help. Please return the form now.

Appendix 3
Map Comparison Formulae

Consider two square adjacency matrices $G = [g_{ij}]$, $H = [h_{ij}]$, where values of both i and j range between 1 and the total number of elements in the two maps, p. The simplest measure (Formula 1, (F1), [6.1]) of the matrix distance between them, $d_{(G,H)}$, as proposed by Langfield-Smith and Wirth (1992) summed the scores of all cells in the distance matrix, the cells of which contain the absolute differences in scores between the corresponding pairs of cells in the two adjacency matrices. Initially, where only causal links between beliefs were considered, individual cells in the adjacency matrix took on values of +1, 0 or -1, defining a maximum distance of 2 in any cell of the distance matrix.

For any two matrices,

$$F1 = d_{(G,H)} = \sum_{j=1}^{p}\sum_{i=1}^{p}|g_{ij} - h_{ij}| \qquad [A3.1]$$

In cases where the cognitive maps of some individuals may include more elements in them than others, measures were proposed [6.2] that weighted the matrix distance in relation to:

1. a maximum distance, based on the number of cells in each matrix which could be occupied;
2. the unoccupied cells on the diagonal;
3. the maximum difference between corresponding cells in G and H.

$$F2 = \sum_{j=1}^{p}\sum_{i=1}^{p}|g_{ij} - h_{ij}| \,/\, 2(p^2 - p) \qquad [A3.2]$$

where, p is the number of elements in the distance matrix and $2(p^2 - p)$ is the maximum distance between the two maps.

As a further refinement, a device to take account of those elements which either, do not impact on, or on which no other elements have an impact, was proposed. A transmitter was defined as an element which has an impact on at least one other element in a map, but on which, no other element has an impact. A receiver is an element which has no impact on any other element, but on which at least one other element has an impact. Given that transmitters form a vertical vector of zeros in an adjacency matrix, and receivers form a horizontal matrix of zeros, corresponding cells in the distance matrix display values of +1, 0 or -1. The total number of transmitters, T, or receivers, R, may either be common to b_o, and R_c respectively, or unique to a particular map, numbering T_u, and R_u respectively. An amended distance measure, Formula 3 (F3), [6.3] then becomes:

$$\text{F3} = \frac{\sum_{j=1}^{p}\sum_{i=1}^{p} | g_{ij} - h_{ij} |}{(p-1)(2p - T - R - T_c - R_c) + T_c.R_c + T.R} \qquad \text{[A3.3]}$$

This in turn can be adjusted (F4, [6.4]) to take account of cases where there are a differing number of elements in each map. This is done by defining p_c, the number of elements common to the two matrices, and p_G the number of unique elements respectively in matrix G and p_H in matrix H.

$$\text{F4} = \frac{\sum_{j=1}^{p}\sum_{i=1}^{p} |g_{ij} - h_{ij}|}{2(p_c^{\,2}) + 2p_c(p_G + p_H) + p_G^{\,2} + p_H^{\,2} - (2p_c + p_G + p_H)} \qquad \text{[A3.4]}$$

When the differing strengths of beliefs are taken into account, raw elements of the adjacency matrix vary from -3 to +3, and when an amended denominator is considered, the distance ratio, Formula 5, (F5), [6.5] becomes:

$$F5 = \frac{\sum_{j=1}^{p}\sum_{i=1}^{p}|g_{ij} - h_{ij}|}{(3p-3)(2p-T-R-T_c-R_c)+3T_c.R_c+3T.R} \qquad [A3.5]$$

This formula still fails to take account of cases where the two maps may contain different sets of elements. An amended formulation, Formula 6 (F6), [6.6] of the denominator was proposed which took account of elements which may be either common to the two maps, yielding a maximum value of +6 in the distance matrix, or unique, with a maximum distance of +1.

$$F6 = \frac{\sum_{j=1}^{p}\sum_{i=1}^{p}|g_{ij} - h_{ij}|}{6(p_c^{\ 2})+2p_c(p_G+p_H)+p_G^{\ 2}+p_H^{\ 2}\text{-}(6p_c+p_G+p_H)} \qquad [A3.6]$$

The matrix distance formula can be broken down [6.7] into those elements which are related to unique beliefs (z), and those which are related to common beliefs. The later includes either those where the strength of commonly held beliefs varies (x), or those where there may or may not be common beliefs with regard to common elements (y).

Thus $\sum_{j=1}^{p}\sum_{i=1}^{p}|g_{ij} - h_{ij}|$ = (x) + (y) + (z).

$$\sum_{j=1}^{p}\sum_{i=1}^{p}|g_{ij} - h_{ij}| = \sum_{\substack{j=1 \\ i,j\,\in pc \\ g_{ij},h_{ij}\neq 0}}^{p}\sum_{i=1}^{p}|g_{ij} - h_{ij}| + \sum_{\substack{j=1 \\ i\,j\,\in pc \\ g_{ij}\text{ or }h_{ij}=0}}^{p}\sum_{i=1}^{p}|g_{ij} - h_{ij}| + \sum_{\substack{j=1 \\ i\text{ or }j\,\notin pc}}^{p}\sum_{i=1}^{p}|g_{ij} - h_{ij}| \qquad [A3.7]$$

Finally, if an attempt is made to adjust for varying strengths of the unique beliefs (y and z) in this formulation, adjustments to the numerator [6.8] are required (Formula 8 (F8)), such that:

$$\text{F8} = \frac{\sum_{j=1}^{p}\sum_{i=1}^{p}|g^*_{ij} - h^*_{ij}|}{6(p_c^{\,2}) + 2p_c(\,p_G + p_H) + p_G^{\,2} + p_H^{\,2} - (6\,p_c + p_G + p_H)} \qquad \text{[A3.8]}$$

where g^*_{ij} = { 1 if $g_j > 0$ and i or $j \notin pc$
{ -1 if $h_j < 0$ and i or $j \notin pc$
{g_{ij} otherwise

and h^*_{ij} is treated likewise.

Appendix 4
Item Scores by Subgroup

Data in Appendices 4.1 to 4.16 show variations in the perceived importance of items in the decision to study amongst subgroups defined by their academic discipline and by cohort.

The percentage of subgroup responses to each item is shown in each cell.

Figures show the percentage of Shipping (Sh.) and Logistics (Lo.) students in each sample of Masters students, by cohort (1997, 1998a, 1998b and 1999).

Source: the author.

Appendix 4.1 The importance of family or friends

How did your family or friends influence your decision to study? How important were they?

% of respondents

Reason		Critical		Important		Relevant		Irrelevant	
		Sh.	Lo.	Sh.	Lo.	Sh.	Lo.	Sh.	Lo.
This factor was an issue for me	1997	0	0	0	0	81	71	19	29
	1998a	13	4	35	37	22	46	30	13
	1998b	33	9	21	41	17	37	29	13
	1999	17	8	38	40	31	36	14	16
My family circumstances are relevant	1997	9	0	10	29	33	24	48	47
	1998a	35	4	13	29	17	21	35	46
	1998b	29	9	22	32	21	37	29	32
	1999	14	8	34	32	14	24	38	36
They introduced me into the industry	1997	0	0	33	0	10	18	57	82
	1998a	17	0	4	4	8	8	61	88
	1998b	21	0	13	0	8	13	58	87
	1999	7	0	24	8	17	12	52	80
I talked to them	1997	5	6	29	35	28	29	38	30
	1998a	26	0	35	50	13	29	26	21
	1998b	25	0	25	39	25	43	25	18
	1999	14	4	41	40	31	48	14	8
They are funding my studies	1997	5	23	9	12	5	12	81	53
	1998a	13	8	4	4	22	25	61	63
	1998b	13	7	8	14	17	23	62	56
	1999	7	8	34	16	21	20	38	56
Other	1997	5	0	0	0	0	0	95	100
	1998a	13	4	4	8	0	0	83	88
	1998b	8	7	4	0	0	0	88	93
	1999	21	20	10	28	14	16	55	36

Appendix 4.2 Postgraduate study of shipping or logistics

What reasons made studying logistics or shipping at postgraduate or post-experience level attractive? How important was each?
% of respondents

Reason		Critical		Important		Relevant		Irrelevant	
		Sh.	Lo.	Sh.	Lo.	Sh.	Lo.	Sh.	Lo.
This factor was	1997	5	0	0	0	95	100	0	0
an issue for me	1998a	48	29	39	58	0	13	13	0
	1998b	42	21	33	73	8	6	17	0
	1999	3	0	4	0	0	4	93	96
I wanted:									
to broaden my	1997	14	29	57	65	24	6	5	0
knowledge /	1998a	39	37	52	50	4	13	5	0
learn new things	1998b	54	25	29	54	9	21	8	0
	1999	31	16	62	68	7	16	0	0
to enact my	1997	33	29	53	53	14	18	0	0
long term career	1998a	56	13	35	58	9	21	0	8
plans	1998b	46	20	33	58	17	22	4	0
	1999	31	36	56	60	10	4	3	0
a change of	1997	19	0	5	23	5	18	71	59
career / to go	1998a	39	0	13	8	9	4	39	88
ashore	1998b	17	0	6	8	4	4	23	88
	1999	41	40	49	36	3	24	7	0
to broaden my	1997	24	29	24	53	28	6	24	12
opportunities /	1998a	39	8	30	54	13	25	18	13
be sure to find a	1998b	38	11	29	60	16	20	17	9
job	1999	24	8	31	12	14	4	31	76
to specialise in	1997	28	18	29	12	24	17	19	53
shipping as I	1998a	48	8	17	4	26	21	9	67
wanted to work	1998b	46	4	25	4	21	6	8	86
there	1999	34	28	42	56	24	12	0	4
more study,	1997	14	6	19	12	24	29	43	53
following my	1998a	13	0	30	29	9	21	48	50
work experience	1998b	13	0	33	12	25	31	29	57
	1999	41	8	31	8	21	28	7	56

Appendix 4.3 What might put you off study?

What reasons might have put you off studying logistics or shipping at postgraduate or post- experience level? How important was each?
% of respondents

		Critical		Important		Relevant		Irrelevant	
Reason		Sh.	Lo.	Sh.	Lo.	Sh.	Lo.	Sh.	Lo.
This factor was an issue for me	1997	0	0	0	0	71	59	29	41
	1998a	26	12	18	13	26	13	30	62
	1998b	25	4	38	20	12	34	25	42
	1999	21	4	41	32	17	40	21	24
I did not yet have the money / was saving up, while at sea	1997	9	0	10	12	19	29	62	59
	1998a	26	8	9	8	4	9	61	75
	1998b	25	4	17	8	8	27	50	61
	1999	24	16	17	24	14	28	45	32
I needed ops. experience to improve my understanding	1997	14	0	5	0	0	6	81	94
	1998a	0	0	9	0	9	8	82	92
	1998b	4	0	21	0	29	13	46	87
	1999	3	0	14	16	28	32	55	52
Had I met work or family business pressures	1997	9	0	0	12	10	0	81	88
	1998a	9	0	4	4	13	0	74	96
	1998b	17	0	4	4	25	8	54	88
	1999	10	0	14	24	17	20	59	56
I had to be sure it was the right course: I could not repeat it	1997	0	0	29	18	0	29	71	53
	1998a	4	8	35	0	9	4	52	88
	1998b	25	0	25	20	13	7	37	73
	1999	17	8	24	24	17	32	42	36
If I was offered a job giving good experience	1997	0	0	33	18	10	6	57	76
	1998a	4	0	22	4	26	0	48	96
	1998b	8	7	13	16	29	16	50	61
	1999	10	16	35	28	24	24	31	32
Returning to study after 10 years working would be hard	1997	0	0	5	23	24	12	71	65
	1998a	4	4	9	0	9	0	78	96
	1998b	12	85	8	11	13	4	67	0
	1999	10	0	4	24	17	20	69	56

Appendix 4.4 Finding out about courses (1)

How did you find out about courses in these areas? How important was each source?

% of respondents

Reason		Critical		Important		Relevant		Irrelevant	
		Sh.	Lo.	Sh.	Lo.	Sh.	Lo.	Sh.	Lo.
This factor was an issue for me	1997	0	0	0	0	100	94	0	6
	1998a	18	13	48	33	30	54	4	0
	1998b	17	12	38	40	33	45	12	3
	1999	24	4	48	48	28	48	0	0
I talked to a work friend / people in industry	1997	19	6	9	17	43	12	29	65
	1998a	26	0	35	4	13	8	26	88
	1998b	8	0	42	0	25	14	25	86
	1999	21	4	31	24	31	16	17	56
My previous lecturers told me about it	1997	14	29	10	41	9	6	67	24
	1998a	4	21	9	38	4	12	83	29
	1998b	8	27	4	20	8	44	80	9
	1999	10	8	24	24	10	32	55	36
I contacted the British Council	1997	5	0	33	0	14	0	48	100
	1998a	26	4	4	0	13	4	57	92
	1998b	17	92	17	4	12	0	54	4
	1999	28	0	0	0	10	16	62	84

Appendix 4.5 Finding out about courses (2)

How did you find out about courses in these areas? How important was each source?

% of respondents

Reason		Critical		Important		Relevant		Irrelevant	
		Sh.	Lo.	Sh.	Lo.	Sh.	Lo.	Sh.	Lo.
I read magazines / books / brochures	1997	10	6	24	18	14	29	52	47
	1998a	9	17	30	12	13	38	48	33
	1998b	8	4	21	37	33	22	38	37
	1999	10	16	28	24	17	32	45	28
By chance	1997	0	0	5	0	9	18	86	82
	1998a	0	4	13	8	0	0	87	88
	1998b	8	4	8	8	13	11	71	77
	1999	0	0	10	20	10	28	80	52
I talked to others planning to study there	1997	5	6	5	18	0	17	90	59
	1998a	4	4	0	42	26	12	70	42
	1998b	8	8	0	25	25	36	67	31
	1999	7	4	17	32	31	16	45	48
I talked to students / former students on the course	1997	10	0	14	12	24	6	52	82
	1998a	22	8	17	37	17	17	44	38
	1998b	25	13	25	13	12	49	38	25
	1999	17	20	24	48	17	8	42	24
Other	1997	9	0	0	0	5	0	86	100
	1998a	17	4	0	4	0	4	83	88
	1998b	17	0	0	0	0	4	83	96
	1999	10	4	4	8	0	0	86	88

Appendix 4.6 Attractions of study in the UK (1)

What reasons made studying logistics or shipping in the UK attractive? How important was each?
% of respondents

Reason		Critical		Important		Relevant		Irrelevant	
		Sh.	Lo.	Sh.	Lo.	Sh.	Lo.	Sh.	Lo.
This factor was an issue for me	1997	0	0	0	0	95	100	5	0
	1998a	26	17	52	63	18	8	4	12
	1998b	33	16	50	54	4	17	13	13
	1999	21	12	62	68	14	20	3	0
I wanted:									
to go abroad as it makes life more interesting	1997	0	18	28	35	24	29	48	18
	1998a	17	13	26	37	18	25	39	25
	1998b	8	11	25	31	21	32	46	26
	1999	21	20	21	52	17	16	41	12
a British Diploma or MSc, recognised around the world	1997	33	29	38	59	10	6	19	6
	1998a	31	21	52	42	13	21	4	16
	1998b	38	26	37	40	21	21	4	13
	1999	28	40	45	36	17	20	10	4
an MSc: we don't have them in my country	1997	14	12	19	23	14	12	53	53
	1998a	22	4	22	4	13	9	43	83
	1998b	21	78	21	7	21	4	37	13
	1999	14	8	24	20	10	28	52	44

Appendix 4.7 Attractions of study in the UK (2)

What reasons made studying logistics or shipping in the UK attractive? How important was each?

% of respondents

		Critical		Important		Relevant		Irrelevant	
Reason		Sh.	Lo.	Sh.	Lo.	Sh.	Lo.	Sh.	Lo.
I wanted:									
to practise	1997	24	29	19	41	29	24	28	6
English / the	1998a	31	29	17	29	13	21	39	21
language of	1998b	29	45	13	21	33	17	25	17
shipping	1999	41	44	28	20	0	32	31	4
a UK course	1997	5	12	10	24	14	23	71	41
which can be	1998a	17	8	18	12	0	13	65	67
completed in a	1998b	4	12	29	16	8	4	59	68
short time	1999	17	20	10	32	28	16	45	32
an academic	1997	9	6	29	6	19	41	43	47
system which is	1998a	17	8	18	17	26	17	39	58
more relevant to	1998b	17	12	37	8	17	30	29	50
industry	1999	24	12	35	28	31	20	10	40
to study in a land	1997	3	0	10	6	48	6	29	88
with 500 years of	1998a	9	0	17	13	26	12	48	75
shipping tradition	1998b	8	4	37	0	17	27	38	69
	1999	17	0	14	12	35	28	34	60

Appendix 4.8 Attractions of study at Plymouth (1)

What reasons made studying Logistics/Shipping at Plymouth attractive?
% of respondents

Reason		Critical		Important		Relevant		Irrelevant	
		Sh.	Lo.	Sh.	Lo.	Sh.	Lo.	Sh.	Lo.
This factor was	1997	0	0	0	0	100	88	0	12
an issue for me	1998a	9	4	52	33	35	29	4	34
	1998b	13	4	33	36	33	38	21	22
	1999	10	8	52	36	31	44	7	12
Plymouth's reputation:									
was	1997	5	6	14	6	19	47	62	41
commended by	1998a	0	0	4	21	26	21	70	58
my lecturers	1998b	4	0	8	17	8	43	80	40
	1999	7	4	20	20	21	28	52	48
is worldwide,	1997	29	6	19	18	28	17	24	59
important when	1998a	13	4	30	17	26	17	31	62
looking for a	1998b	12	4	38	4	4	17	46	75
job	1999	14	4	48	28	17	36	21	32
The Plymouth course:									
is the only one	1997	9	0	14	6	10	23	67	71
which offers an	1998a	0	0	13	4	13	4	74	92
MSc	1998b	0	0	17	0	29	7	54	93
	1999	3	4	10	8	11	20	76	68
is the only one	1997	9	18	0	23	5	18	86	41
in International	1998a	0	4	4	8	4	21	92	67
Logistics	1998b	0	8	0	13	12	29	88	50
	1999	0	8	0	8	7	44	93	40
specialises in	1997	33	0	43	12	5	0	19	88
shipping	1998a	26	0	52	17	13	8	9	75
	1998b	37	0	38	8	17	21	8	71
	1999	31	8	45	20	10	16	14	56
offers an MSc:	1997	10	0	14	18	14	17	62	65
the Diploma is	1998a	22	4	17	8	4	4	57	84
an entry to this	1998b	21	3	13	0	25	11	41	86
	1999	10	4	28	16	7	20	55	60
could give me	1997	9	0	10	6	10	6	71	88
remission from	1998a	9	0	22	4	13	4	56	92
MSc Stage 1	1998b	17	3	21	0	8	3	54	94
	1999	10	0	17	0	4	12	69	88

Appendix 4.9 Attractions of study at Plymouth (2)

What reasons made studying logistics/shipping at Plymouth attractive?
% of respondents

		Critical		Important		Relevant		Irrelevant	
In the city of Plymouth:		Sh.	Lo.	Sh.	Lo.	Sh.	Lo.	Sh.	Lo.
I have friends	1997	0	0	19	6	10	0	71	94
who have	1998a	4	0	0	8	18	8	78	84
finished the	1998b	0	3	12	3	25	22	63	72
course	1999	3	0	7	16	17	12	72	72
I have already	1997	5	0	9	0	24	6	62	94
undertaken	1998a	4	8	4	13	9	4	83	75
studies	1998b	4	8	8	8	12	13	76	71
	1999	10	4	7	8	24	12	59	76
I would live in a	1997	0	0	19	6	14	6	67	88
nice place /	1998a	0	4	17	4	22	17	61	75
beautiful city	1998b	4	4	21	17	29	17	46	62
	1999	7	0	7	12	34	40	52	48
Plymouth University:									
has a good	1997	10	0	9	12	33	18	48	70
admissions	1998a	4	0	17	8	18	4	61	88
administration	1998b	4	0	29	8	13	12	54	80
	1999	14	4	21	16	38	32	28	48
accepted me	1997	9	6	24	17	38	6	29	71
	1998a	39	8	17	8	31	17	13	67
	1998b	29	8	42	13	17	33	12	46
	1999	31	12	34	28	21	28	14	32
did not have bad	1997	14	0	10	18	19	0	57	82
oral reports like	1998a	0	4	26	8	26	4	48	84
other places	1998b	8	4	32	4	8	15	52	77
	1999	7	4	17	8	35	32	41	56
lecturers have a	1997	19	6	19	12	38	6	24	76
good reputation	1998a	9	4	26	17	22	8	43	71
	1998b	8	0	42	16	13	30	37	54
	1999	10	8	41	4	28	36	21	52

Appendix 4.10 Attractions of study at Plymouth / St Nazaire (3)

What reasons made studying logistics or shipping at Plymouth (or St Nazaire / Plymouth for Eurodip students) attractive? How important was each?

% of respondents

Reason		Critical		Important		Relevant		Irrelevant	
		Sh.	Lo.	Sh.	Lo.	Sh.	Lo.	Sh.	Lo.
For St Nazaire students only: by choosing St Nazaire, its links with Plymouth gave me:									
a chance to study	1997	0	0	5	0	0	6	95	94
in Plymouth,	1998a	0	0	0	0	0	0	100	100
and not just any	1998b	0	0	0	0	0	0	100	100
UK city	1999	0	0	0	0	0	0	100	100
a recognised	1997	0	0	0	0	0	0	100	100
degree, which	1998a	0	0	0	0	0	0	100	100
makes it easier	1998b	0	0	0	0	0	4	100	96
to find a job	1999	0	0	0	0	0	0	100	100
access to more	1997	0	0	5	6	0	0	95	94
study facilities	1998a	0	0	0	0	4	0	96	100
	1998b	0	0	0	0	4	0	96	100
	1999	0	0	0	0	0	0	100	100

Appendix 4.11 Reasons which might have put me off Plymouth

What reasons might have put you off studying logistics or shipping at Plymouth? How important was each?

% of respondents

Reason		Critical		Important		Relevant		Irrelevant	
		Sh.	Lo.	Sh.	Lo.	Sh.	Lo.	Sh.	Lo.
This factor was an issue	1997	0	0	5	0	67	82	28	18
	1998a	26	4	39	29	4	29	31	38
	1998b	33	27	29	25	13	44	25	4
	1999	17	8	48	52	28	28	7	12
If I had found that Plymouth University:									
gave me too little information	1997	5	12	28	29	24	24	43	35
	1998a	13	4	35	9	13	33	39	54
	1998b	25	0	33	33	21	19	21	48
	1999	24	8	49	48	24	32	3	12
was costly in relation to other courses	1997	5	0	29	29	19	30	47	41
	1998a	22	0	17	17	9	17	52	66
	1998b	21	4	33	8	8	40	38	48
	1999	17	12	21	36	41	24	21	28
rated low on course quality, library	1997	14	12	29	35	19	6	38	47
	1998a	30	4	26	13	0	21	44	62
	1998b	33	4	38	21	12	31	17	44
	1999	28	12	45	32	17	36	10	20
had not offered me a place at MSc level	1997	38	41	10	12	19	18	33	29
	1998a	44	17	13	25	4	0	39	58
	1998b	50	12	17	22	17	16	16	50
	1999	66	54	14	25	3	13	17	8
qualifications were not unique	1997	5	18	14	12	24	41	57	29
	1998a	9	0	35	12	8	21	48	67
	1998b	21	0	29	12	21	17	29	71
	1999	21	4	52	16	17	48	10	32

Appendix 4.12 Reasons which might have put me off logistics / shipping

What reasons might have put you off studying logistics or shipping at Plymouth? How important was each?

% of respondents

Reason		Critical		Important		Relevant		Irrelevant	
		Sh.	Lo.	Sh.	Lo.	Sh.	Lo.	Sh.	Lo.
If I had found that Plymouth University:									
considered my grades to be inadequate	1997	0	18	33	23	19	24	48	35
	1998a	13	8	35	13	8	0	44	79
	1998b	21	4	42	12	12	4	25	80
	1999	31	20	34	24	14	32	21	24
was not the best school which I could afford	1997	10	12	9	6	24	23	57	59
	1998a	26	0	13	4	9	13	52	83
	1998b	25	8	29	4	4	16	42	72
	1999	17	8	34	24	28	24	21	44
students had given me bad reports about it	1997	14	0	19	18	29	29	38	53
	1998a	4	4	30	29	18	17	48	50
	1998b	25	8	42	37	12	16	21	39
	1999	24	8	35	40	24	24	17	28
Other	1997	5	0	0	6	0	6	95	88
	1998a	4	0	0	4	0	0	96	96
	1998b	4	8	0	4	0	0	96	88
	1999	7	4	0	4	0	0	93	92

Appendix 4.13 Reasons for not studying elsewhere

Think of courses that you could have studied elsewhere. Why did you not study elsewhere? How important was each reason?

% of respondents

		Critical		Important		Relevant		Irrelevant	
Reason		Sh.	Lo.	Sh.	Lo.	Sh.	Lo.	Sh.	Lo.
This factor was	1997	0	0	0	0	67	71	33	29
an issue for me	1998a	22	17	26	25	35	29	17	29
	1998b	17	4	37	36	25	21	21	39
	1999	14	20	38	40	31	32	17	8
Elsewhere:									
it was a totally	1997		6	10	17	14	6	57	71
different subject	1998a	22	17	22	8	17	8	39	67
	1998b	12	12	25	4	25	30	38	54
	1999	17	16	28	28	28	28	27	28
I would have	1997	10	0	5	18	9	17	76	65
had to repeat	1998a	0	4	22	8	17	25	61	63
material	1998b	4	4	21	8	13	40	62	48
	1999	10	8	28	24	10	32	52	36
the course	1997	19	0	19	12	24	12	38	76
emphasis was	1998a	13	4	13	17	35	17	39	62
wrong for me	1998b	25	0	21	12	17	30	37	58
	1999	14	12	34	24	24	32	28	32
it was very	1997	14	17	10	12	14	6	62	65
expensive, and	1998a	13	12	22	13	17	21	48	54
not necessarily	1998b	21	4	33	26	17	21	29	49
better	1999	14	8	17	20	21	28	48	44
the course took	1997	5	12	5	17	5	12	85	59
longer to	1998a	22	17	21	21	22	8	35	54
complete	1998b	25	8	21	25	21	27	33	40
	1999	7	28	21	32	10	12	62	28

Appendix 4.14 Barriers to study (1)

How important is each of the following barriers that might have put you off studying logistics or shipping at Plymouth?
% of respondents

		Critical		Important		Relevant		Irrelevant	
Reason		Sh.	Lo.	Sh.	Lo.	Sh.	Lo.	Sh.	Lo.
This factor overall was an issue for me	1997	0	0	0	0	81	94	19	6
	1998a	4	4	35	13	39	54	22	29
	1998b	8	4	42	18	25	44	25	34
	1999	7	16	48	28	35	48	10	8
Money:									
This factor was an issue for me	1997	0	0	0	0	43	59	57	41
	1998a	13	8	26	17	26	42	35	33
	1998b	13	8	33	12	21	36	33	44
	1999	10	20	34	40	21	28	35	12
I was willing to pay a bit more for what I wanted	1997	0	0	19	6	10	17	71	77
	1998a	4	0	18	4	30	25	48	71
	1998b	21	0	5	0	37	18	37	82
	1999	3	4	21	28	28	36	48	32
The total cost was less, when I considered the time involved	1997	0	6	5	6	9	18	86	70
	1998a	4	4	13	8	18	17	65	71
	1998b	12	4	17	0	21	24	50	74
	1999	0	0	21	12	24	32	55	56
My company or parents partly funded / encouraged me	1997	19	23	5	24	19	12	57	41
	1998a	22	13	13	21	22	17	43	50
	1998b	25	12	8	4	8	32	59	52
	1999	24	16	24	28	14	28	38	28
	1999	3	4	4	4	0	0	93	92
The quality of life in the city, my housing etc.	1997	0	0	38	30	19	35	43	35
	1998a	8	0	39	12	17	38	35	50
	1998b	12	0	38	12	17	35	33	53
	1999	7	0	38	8	31	68	24	24

Appendix 4.15 Barriers to study (2)

How important is each of the following barriers that might have put you off studying logistics or shipping at Plymouth?

% of respondents

Reason		Critical		Important		Relevant		Irrelevant	
		Sh.	Lo.	Sh.	Lo.	Sh.	Lo.	Sh.	Lo.
Teaching methods on the course:									
This factor was an issue for me	1997	0	0	0	0	86	83	14	17
	1998a	26	0	26	29	39	8	9	63
	1998b	37	4	17	16	33	30	13	50
	1999	7	8	65	44	21	32	7	16
My first contacts with staff made me feel welcome	1997	19	6	33	35	29	41	19	18
	1998a	22	4	30	21	39	13	9	62
	1998b	17	8	33	12	33	30	17	50
	1999	21	8	45	52	17	28	17	12
I expected to understand, not to analyse statistics	1997	19	12	29	23	24	41	28	24
	1998a	30	0	44	25	9	13	17	62
	1998b	38	4	33	36	21	8	8	52
	1999	21	8	28	40	41	36	10	16
My practical experience was under-valued	1997	0	6	9	12	5	23	86	59
	1998a	9	4	9	0	26	21	56	75
	1998b	17	0	21	0	29	12	33	88
	1999	0	4	21	4	27	12	52	80
I expected visits to companies and professionals	1997	14	0	19	41	19	29	48	30
	1998a	30	4	35	12	9	17	26	67
	1998b	46	8	25	4	12	32	17	56
	1999	17	12	52	28	21	40	10	20

Appendix 4.16 Other issues

Are there any issues that have been overlooked?
% of respondents

These reasons were:		Critical		Important		Relevant		Irrelevant	
Reason		Sh.	Lo.	Sh.	Lo.	Sh.	Lo.	Sh.	Lo.
Issue 1	1997	14	18	0	0	0	6	86	76
	1998a	22	8	0	0	0	4	78	88
	1998b	17	12	8	0	0	4	75	84
	1999	10	0	7	8	0	4	83	88
Issue 2	1997	5	6	0	0	0	0	95	94
	1998a	0	0	13	4	0	0	87	96
	1998b	8	4	4	4	0	0	88	92
	1999	4	0	0	0	0	4	96	96

Appendix 4.17 Respondents profile by work experience

Work experience relevant to the course by location of work experience.
% of respondents

	Seagoing		Port		Inland		None	
	Ship	Log	Ship	Log	Ship	Log	Ship	Log
1997	33	6	38	17	10	65	19	12
1998a	39	4	9	13	48	62	4	21
1999	52	12	7	12	27	68	14	8

Full / part time experience

	Full time		Part time		None	
	Ship	Log	Ship	Log	Ship	Log
1997	67	88	14	0	19	12
1998a	70	58	26	21	4	21
1999	55	72	31	20	14	8

Years of work experience

	Under 1		1-3		Over 3		None	
	Ship	Log	Ship	Log	Ship	Log	Ship	Log
1997	34	70	27	0	20	18	19	12
1998a	4	22	39	39	53	18	4	21
1999	10	36	42	36	31	20	17	8

Appendix 4.18 Respondent profile by gender, age, nationality, work experience

Respondent profile: the % of respondents in each group

Sample Gender		Male		Female	
		Ship	Log	Ship	Log
	1997	81	65	19	35
	1998	96	92	4	8
	1999	86	76	14	24

Sample ages:		20-29		30-39		40-49	
		Ship	Log	Ship	Log	Ship	Log
	1997	90	82	10	12	0	6
	1998a	79	79	17	17	4	4
	1999	79	92	14	8	7	0

Nationality	1997		1998		1999	
	Ship	Log	Ship	Log	Ship	Log
United Kingdom	14	0	5	12	3	0
France	5	0	0	0	0	12
North Europe	28	88	13	63	17	64
South Europe	33	6	30	8	55	20
Africa	5	6	13	0	0	0
Asia	10	0	26	13	7	4
North America	5	0	4	0	10	0
South America	0	0	0	4	3	0
Other	0	0	9	0	3	0

For Product Safety Concerns and Information please contact our EU representative GPSR@taylorandfrancis.com Taylor & Francis Verlag GmbH, Kaufingerstraße 24, 80331 München, Germany

Printed and bound by CPI Group (UK) Ltd, Croydon, CR0 4YY
01/05/2025
01858351-0004